T0224291

# Deploying AI in the Enterprise

IT Approaches for Design, DevOps, Governance, Change Management, Blockchain, and Quantum Computing

Eberhard Hechler
Martin Oberhofer
Thomas Schaeck

*Foreword by Srinivas Thummalapalli*

Apress®

*Deploying AI in the Enterprise: IT Approaches for Design, DevOps, Governance, Change Management, Blockchain, and Quantum Computing*

Eberhard Hechler
Sindelfingen, Germany

Martin Oberhofer
Boeblingen, Germany

Thomas Schaeck
Boeblingen, Germany

ISBN-13 (pbk): 978-1-4842-6205-4
https://doi.org/10.1007/978-1-4842-6206-1

ISBN-13 (electronic): 978-1-4842-6206-1

Managing Director, Apress Media LLC: Welmoed Spahr
Acquisitions Editor: Joan Murray
Development Editor: Laura Berendson
Coordinating Editor: Jill Balzano

Cover image designed by Freepik (www.freepik.com)

Distributed to the book trade worldwide by Springer Science+Business Media New York, 233 Spring Street, 6th Floor, New York, NY 10013. Phone 1-800-SPRINGER, fax (201) 348-4505, e-mail orders-ny@springer-sbm.com, or visit www.springeronline.com. Apress Media, LLC is a California LLC and the sole member (owner) is Springer Science + Business Media Finance Inc (SSBM Finance Inc). SSBM Finance Inc is a **Delaware** corporation.

For information on translations, please e-mail booktranslations@springernature.com; for reprint, paperback, or audio rights, please e-mail bookpermissions@springernature.com.

Apress titles may be purchased in bulk for academic, corporate, or promotional use. eBook versions and licenses are also available for most titles. For more information, reference our Print and eBook Bulk Sales web page at http://www.apress.com/bulk-sales.

Any source code or other supplementary material referenced by the author in this book is available to readers on GitHub via the book's product page, located at www.apress.com/9781484262054. For more detailed information, please visit http://www.apress.com/source-code.

Printed on acid-free paper

*To my wife, Irina, and our two sons, Lars and Alex, for their continuing support and understanding in writing this book on long evenings and weekends instead of spending time with them.*

—*Eberhard Hechler*

*To my wife, Kirsten, and our two sons, Damian and Adrian, thank you for all the love and inspiration you give me every day.*

—*Martin Oberhofer*

*To my wife, Annette, and our children, Amelie and Felix, for their support and patience while I was contributing to this book.*

—*Thomas Schaeck*

# Table of Contents

# About the Authors

**Eberhard Hechler** is an Executive Architect at the IBM Germany R&D Lab. He is a member of the Db2 Analytics Accelerator development group and addresses the broader data and AI on IBM Z scope, including machine learning for z/OS. After 2.5 years at the IBM Kingston Lab in New York, he worked in software development, performance optimization, IT/solution architecture and design, open source (Hadoop and Spark) integration, and master data management (MDM).

He began to work with Db2 for MVS, focusing on testing and performance measurements. He has worked worldwide with IBM clients from various industries on a vast number of topics, such as data and AI including analytics and machine learning, information architectures (IA), and industry solutions. From 2011 to 2014, he was at IBM Singapore, working as Lead Big Data Architect in the Communications Sector of IBM's Software Group.

Eberhard has studied in Germany and France and holds a master's degree (Dipl.-Math.) in pure mathematics and a bachelor's degree (Dipl.-Ing. (FH)) in electrical engineering. He is a member of the IBM Academy of Technology Leadership Team and coauthored the following books: *Enterprise MDM*, *The Art of Enterprise Information Architecture*, and *Beyond Big Data*.

**Martin Oberhofer** is an IBM Distinguished Engineer and Executive Architect. He is a technologist and engineering leader with deep expertise in master data management, data governance, data integration, metadata and reference data management, artificial intelligence, and machine learning. He has a proven track record of translating customer needs into software solutions, working collaboratively with globally distributed development, design, and offering management teams. He guides development teams using Agile and DevOps software development methods. He can easily adapt to ever-present challenges. Recently, he also started to dig into the blockchain technology space, exploring opportunities to bring analytics capabilities to the blockchain realm.

Previous to his current assignment in the IBM Data and AI development organization, Martin worked with many large clients worldwide at the enterprise level, providing thought leadership on data-centric solutions. In this role, he demonstrated his ability to think horizontally to bring business and IT together by communicating solutions to complex problems in simple terms.

He is an elected member of the IBM Academy of Technology and the TEC CR. He is a certified IBM Master Inventor with over 100 granted patents and numerous publications, including 4 books.

**Thomas Schaeck** is an IBM Distinguished Engineer (technical executive) at IBM Data and AI, leading Watson Studio on IBM Cloud (Cloud Pak for Data) Desktop and integration with other IBM offerings. Watson Studio is a cloud-native collaborative data science and AI environment for data scientists, data engineers, AI experts, business analysts, and developers, allowing teams to gain insights, train, define and deploy ML/DO models, and get from insights to optimal actions. Previously, Thomas led architecture and technical strategy for IBM Connections, WebSphere Portal, and IBM OpenPages. On a 1-year assignment in the USA in 2013–2014, Thomas led transformation of architecture, technical strategy, and DevOps process for IBM OpenPages Governance Risk Compliance, drove adoption of IBM Design Thinking, and became a trusted partner for major IBM OpenPages customers.

Previously, Thomas led architecture and technical strategy for IBM Connections and integration with WebSphere Portal, enterprise content management, business process management, and design and development of Smart Social Q&A, became a trusted partner for large-enterprise customers as well as customer councils, and helped accelerate sales. On a 2-year assignment in the USA in 2004–2006, Thomas led collaboration software architecture, development, and performance for messaging and web conferencing, achieving acceleration of development productivity and large improvements in performance and scalability.

Thomas also led architecture and technical direction for WebSphere Portal Platform and development of the WebSphere Portal Foundation, initiated and led the portal standards Java Portlet API and OASIS WSRP and Apache open source reference implementations, and initiated and led the Web 2.0 initiative for WebSphere Portal. As a trusted portal architect and leader in portal integration standards, he played a key role in winning the hearts and minds of initial reference customers and then many enterprise customers in Germany and Europe.

# About the Technical Reviewer

**Mike Sherman** has over 35 years of marketing, market research, and CRM/big data experience. He helps clients address marketing opportunities through understanding end users' needs, turning them into insight/data specifications, and converting the output into clear, actionable results.

Mike recently published his first (and last!) book, *52 Things We Wish Someone Had Told Us About Customer Analytics*, coauthored with his son Alex. The book captures real-life lessons learned over their careers, with a focus on practical applications of analytics that connect methodologies and processes to impactful outcomes.

Mike began his career at Procter & Gamble, where he managed both new and established brands. Mike spent 17 years with McKinsey & Company; while there, he created their Asia-Pacific marketing practice and was a founder of their global CRM practice. Mike was also Global Head of Knowledge Management for Synovate, where he led efforts to improve the value clients obtain from research. At SingTel and Hong Kong Telecom, he set up their big data teams and drove the use of both customer data and customer research to help the business understand customer and customer data opportunities.

Mike has been based in Asia since 1997 and has supported work in almost every country in the Asia-Pacific region. Mike has extensive experience in the telecom, retail, financial services, consumer electronics, and FMCG industries.

Mike has an MBA, High Distinction (Baker Scholar) from Harvard Business School, and two bachelor's degrees, magna cum laude, from the Wharton School and College, University of Pennsylvania.

Mike is a frequent speaker at conferences and published several times in the McKinsey Quarterly on marketing issues in developing Asian markets. He is the former Board Chair of AFS-USA, a leading high-school foreign exchange organization, and an avid traveler, having visited over 140 countries.

# Foreword

Artificial intelligence is a broad term that has captured people's interest. Until recently, it was confined to scientific circles as one of the most advanced branches of study and struggled to find its way into the industrial arena. While many reasons can be cited for its delayed entry, I strongly believe it is due to the lack of prescriptive books, like the one you hold in your hand. It is the ever-increasing speed of processors churning out data in massive volume that necessitated automation in the data space. AI precisely handles this task of "understanding the data" while producing analytical results. With increasing volumes of data, industries were frantically searching for new tools to analyze the data, and AI came to the rescue at the right time. However, AI is complex enough that only a few in the mainstream can make sense of it. This author team comes to the rescue at the right moment, by providing insight with great examples.

AI has started showing promise recently in the commercial space as many have started using AI for simple use cases to eliminate some mundane tasks that can be automated. From a very early age of my childhood, I was fascinated with the scientific approach to solving problems and the curiosity only grew with time for me. AI grabbed my attention when I was looking for ways to address various problems the financial industry was trying to solve. At that time, I had an opportunity to design systems as a chief enterprise architect, positioning me to merge my curiosity with my role at work. I have been following AI since then and looking for ways to use it in the industrial space. There have been many use cases that came to reality recently using AI. The prompt-based phone answering systems we used to use are now being replaced by speech-enabled systems, empowering customers to directly ask for what they want rather than the system taking them (painfully) through prompts. The Internet is now fully sprinkled with chatbots to take over customer service online. The financial industry has long been using machine learning (a branch of AI) for predictive data analytics and automated decision making in various applications. Tesla and many other firms have started using AI in autonomous vehicles. Increasingly, many companies have started exploring AI/ML for potential use cases in their space.

AI/ML is a complex subject and requires well-versed authors to introduce it to a widespread audience. In this book, Eberhard, Martin, and Thomas did a great job of introducing the subject in plain English. They have successfully bridged the gap between AI and the mainstream audience and explained the subject in simple terms. This book is appropriate for audiences ranging from enthusiastic readers to scientific researchers. It will help people in various roles, such as analysts, programmers, architects, business managers, senior managers, and C-level executives, to be exposed to the subject. The book will bring its audience from a level of having no knowledge about AI/ML to a good understanding of the field, while creating the ability for people to use the tools in their field.

The authors take the reader into the topic of creating an information architecture (IA) surrounding AI/ML. The goal is to create a structure for deploying AI/ML in any organization and helping business executives to absorb it in their own organization. The book very cleanly outlines imperatives for architecture surrounding AI, even to a novice reader. The authors have been able to successfully identify various entities in IA where there is none presently available in the industry in the field of AI. The book will help appropriate personnel in proper positions achieve organizational success.

After IA on AI, the reader is guided through the process of operationalizing AI instead of just being left with an understanding of the subject. Understanding and implementation are two entirely different aspects which the authors clearly understand. The reader is not left with unanswered questions about how to operationalize AI. The authors clearly know that it becomes more relevant in AI than in other fields and have helped to guide the reader in the process. To that end, they have further classified different aspects of AI into subdomains with simple explanations. I find it very important for any level of AI expert to take a concept to implementation.

The authors not only focus on bringing AI/ML into the mainstream but also look at the big picture while correlating AI/ML to other fields such as blockchain and quantum computing. They clearly demonstrate a broad understanding while bringing AI/ML to IT fields such as governance, change management, and DevOps. As AI is a new field and still in experimental stages in the commercial arena, it's not easy to put controls around it. However, the authors understand the importance of governance in the field of IT and don't leave AI without it. They explain the basic need of it in IT before drawing the reader into AI governance itself. The authors have covered the entire scope of AI in language that can be easily understood.

The authors not only highlight AI/ML for their benefits but also show the limitations of the field and suggest new advances required to push the envelope. The book will help professionals prepare for future advances in their field while they are deploying AI at their organization. I personally like this aspect of illustrating the limitations of AI. It shows the depth of an author in a given field to draw out the limitations of the field. Only a well-rounded expert in the field is capable of providing limitations of the field, and the authors no doubt show their depth. I would encourage everyone in IT and scientific fields to read through this book to get an understanding of the field from a fresh set of eyes and to develop a new perspective. The book is a very good read, even for those who are not in technical fields, as it is written in simple English and is in reach to a casual, interested reader of AI. It will improve understanding of the field while enriching one's capability to invite AI/ML into their organization, regardless of the industry.

I congratulate the authors on their well-written book and encourage them to continue to provide valuable bridges and insights in the future.

Srinivas Thummalapalli
Chief Enterprise Architect
Fifth Third Bank
July 2020

# Acknowledgments

Writing a book is a much harder endeavor than we thought and more rewarding than we could have imagined. It requires subject-matter expertise and insight, but also motivation and inspiration. Staying engaged, driving the project forward, improving the chapters, making them more readable, and finding new motivation somewhere weren't always so easy. But now it's done.

We are eternally grateful to the many IBM colleagues, domain experts, and leaders we have worked with around the globe. Collaborating with universities provided us with an invaluable and product-agnostic view regarding artificial intelligence (AI) research topics. Numerous enterprises and organizations that we have had an opportunity to work with in the recent years have provided us with the inspiration and insight in elaborating on some of the AI challenges – and coming up with ideas – in deploying AI into the enterprise.

A very special thanks to *Stephane Rodet*, the Lead UX Engineer from the IBM Germany R&D Lab, who has helped us so much in getting the figures of this book into an attractive and consumable form.

Last but not least, thanks to everyone on the Apress team who helped us so much. Special thanks to *Joan Murray*, the ever-patient acquisitions editor, and *Jill Balzano*, our amazing coordinating editor, the greatest cover designer we could ever imagine.

# Book Layout

This book is for a reader who is looking for guidance and recommendations on how to overcome AI solution deployment and operationalization challenges in an enterprise and is, furthermore, eagerly interested in getting a comprehensive overview on how AI impacts other areas, such as design thinking, information architecture, DevOps, blockchain, and quantum computing – to name a few. The anticipated reader is looking for examples on how to leverage data to derive to actionable insight and predictions and tries to understand current risks and limitations of AI and what this means in an industry-relevant context. We are aiming at IT and business leaders, IT professionals, data scientists, software architects, and readers who have a general interest in getting a holistic AI understanding.

The chapters of this book are organized into four main parts.

**Part I: Getting Started** sets the scene for the book in terms of providing a short introductory chapter, an AI evolutionary perspective including technological advancements, and a short description of the most important AI aspects with machine learning (ML) and deep learning (DL) concepts.

It consists of the following three chapters:

- **Chapter 1: AI Introduction** gives an overview of AI in enterprises, providing examples of high-value use cases and showing how AI can be applied in practice. It describes how to increase enterprise automation using AI and introduces the AI life cycle from an enterprise point of view.

- **Chapter 2: AI Historical Perspective** describes why the theoretical AI underpinning of the second half of the twentieth century led to the remarkable AI boost in the last decade. We also venture a glimpse into the future, briefly elaborating on technological advancements that we will most likely observe in the near future.

- **Chapter 3: Key ML, DL, and DO Concepts** introduces key concepts of ML and decision optimization (DO) and explains the differences between these two concepts. We also discuss labeling data in smart ways to minimize labor cost and expert time in labeling and introduce the concept of automated creation of AI models.

**Part II: AI Deployment** concentrates on successful AI deployments by advocating the implementation of a pervasive information architecture for AI, which is an essential component of every AI deployment that is all too often neglected. We are then providing examples how to turn data into actionable predictions and insight, describing how to leverage ML-based matching for improved and trusted core information management, and sharing guidelines with the reader to overcome operationalization challenges in enterprise environments, including key design thinking and DevOps aspects in the context of AI.

It consists of the following four chapters:

- **Chapter 4: AI Information Architecture** elaborates on the role of an information architecture to deliver a trusted and enterprise-level AI foundation. This chapter is important to the reader in order to fully understand the impact of AI on an existing information architecture to deploy sustainable AI solutions.

- **Chapter 5: From Data to Predictions to Optimal Actions** explains how predictions from ML and decision optimization can be combined to achieve optimal outcomes for enterprises, including a set of practical examples.

- **Chapter 6: The Operationalization of AI** deals with the implementation of AI artifacts into an often highly complex and diverse enterprise environment. This includes real-time scoring; monitoring of, for instance, ML models to maintain their accuracy and precision; and turning data into actionable insight.

- **Chapter 7: Design Thinking and DevOps in the AI Context** describes how design thinking and DevOps methods can be applied to develop AI systems and devices, products and tools, and applications. We also elaborate on how AI and its siblings can be leveraged and infused into design thinking and DevOps concepts.

**Part III: AI in Context** takes into consideration that AI doesn't stand by itself, it exists within a larger context. This third part describes AI in the context of other key initiatives across industries, such as blockchain, quantum computing, governance and master data management, and change management.

It consists of the following five chapters:

- **Chapter 8: AI and Governance** describes AI and governance aspects and, furthermore, discusses the need for explainability, fairness, and traceability. Since AI-based decision making ought to be meaningful and human comprehensible, AI comes with a new dimension of governance imperatives designed to ensure transparency, trust, and accountability.

- **Chapter 9: Applying AI to Data Governance and MDM** provides a deep dive into applying ML to master data management (MDM) and data governance solutions. It specifically highlights the application of ML to improve required matching algorithms for MDM and to discover hidden relationships in core enterprise information.

- **Chapter 10: AI and Change Management** sheds some light on change management in the context of AI and introduces key aspects of AI change management, such as identifying and analyzing sentiments for a more targeted change management with an optimized outcome.

- **Chapter 11: AI and Blockchain** describes the applicability of AI to blockchain, which by itself is still a relatively new concept, and provides examples, for instance, to increase tamperproof audit trail for AI model versions, data sets used in training, and many others.

- **Chapter 12: AI and Quantum Computing** looks at some AI problems, which are likely to benefit from quantum computing. The promise of quantum computing to surpass "classical" computers for some computational problems may have a profound impact on solving AI problems, for instance, complex back-propagation algorithms used to learn high-dimensional artificial neural networks (ANNs).

**Part IV: AI Limitations and Future Challenges** discusses current AI limitations and challenges, some of which are subject to research, while others may constitute insoluble challenges that will leave room for human beings to fill that gap – even in the distant future. Some closing remarks and an outlook into the future of AI will conclude this final part of the book.

It consists of the following two chapters:

- **Chapter 13: Limitations of AI** addresses the promise of AI with its breathtaking range of applications that seem to be without limits. And yet, even for AI, there are a number of limits and future challenges, as we learn about in this chapter.

- **Chapter 14: In Summary and Onward** gives an outlook on likely future evolution of AI and AI adoption and shares thoughts on possible consequences.

# PART I

# Getting Started

# CHAPTER 1

# AI Introduction

Artificial intelligence (AI) has been a vision of humans for a long time. Works of fiction have explored the topic of AI from many angles. For instance, *Neuromancer, 2001: A Space Odyssey, Terminator, A.I., Star Trek, Alien, Mother,* and so forth feature AI in many different manifestations: some human-like and some very different, some serving, some working with, and some even fighting against humans.

While artificial general intelligence (AGI) as featured in science fiction and movies remains more than elusive, there has been a lot of progress in several practical fields of AI which have already moved from fiction to reality.

Especially, the AI areas of machine learning (ML) and deep learning (DL) have advanced from research to practice and are meanwhile applied by a large number of companies and organizations to a stunning breadth of use cases around the world. We are now at a point where leveraging ML and DL is state of the art for modern enterprises, yet larger-scale adoption still lies ahead. Early adopters will go deeper and broader in their ML and DL applications, and those who did not yet start in earnest will need to follow soon.

## AI for the Enterprise

This book provides you with recommendations and best practices to apply AI holistically in an enterprise and organizational context. To help leverage AI and deliver on its promise in a meaningful and business-relevant way, we offer you a pragmatic view on AI and how to unleash its transformational, disruptive power.

© Eberhard Hechler, Martin Oberhofer, Thomas Schaeck 2020
E. Hechler et al., *Deploying AI in the Enterprise*, https://doi.org/10.1007/978-1-4842-6206-1_1

Enterprise AI entails leveraging not only advanced ML and DL but also natural language processing (NLP) and decision optimization (DO) to come to automated actions, robotics, and other areas to optimize existing business processes and to implement new use cases. AI in the enterprise[1] aims to discover organizational knowledge and deliver and infuse analytical insight into decision processes in a way that is as aligned with how a human person would go about these tasks, but accelerates these processes by orders of magnitude.

We provide you with a pervasive view of AI that is driven by challenges and gaps – and subsequently opportunities – for you to gain competitive advantage. One particular area is related to AI life cycle and deployments, including AI operationalization challenges, the need for a comprehensive information architecture (IA) to enable AI by providing the data that fuels it, DevOps aspects, and how to come to actionable decisions based on insight derived from ML/DL and DO models. We also explore AI in the context of specific areas, such as master data management (MDM), governance and change management, and blockchain, as well as future directions such as quantum computing.

The applicability of AI for the enterprise[2] offers a rather diverse perspective. ML, DL, and DO are key areas that we stay focused on throughout the entire book. In this introduction and furthermore in Chapter 5, "*From Data to Predictions to Optimal Actions*," we give you an understanding of the complementary nature of ML/DL on the one hand and DO on the other hand, allowing to get from data to predictions to optimized decisions to enable automated actions. In addition, we also provide you with a high-level description of the progression and advancement of AI in the last few decades, as this is essential in order to appreciate the maturity – or rather missing capabilities – of AI.

A discussion of AI in the enterprise requires an intensive treatment of an AI information architecture and the challenging operationalization aspects of AI, including DevOps in the context of AI. No enterprise can sustain today's business dynamics and required business agility without a robust information architecture (IA), including aspects like data storage and management, governance and change management, and master data management (MDM).

The impact of AI and the opportunities that AI represents for these areas is extensively discussed in Part 3, *AI in Context*.

---

[1]See [1] and [2] for more information on AI in the enterprise.

[2]See [3] for more information on the AI-powered enterprise.

# AI Objective: Automated Actions

What most companies and organizations ultimately are looking to get out of applying AI are automated decisions to drive automated actions in order to accelerate their business or other objectives or assisting human decisions with recommendations where it is not possible or not acceptable to replace human judgment. Automating or assisting decisions and resulting actions can lead to enormous efficiencies and speed and in some cases can enable entirely new business models that would otherwise be impossible – for example, modern ecommerce, fraud detection, dating apps, and many more.

On the flip side, if not done well, automation of decisions and actions can lead to damage or losses – such as a self-driving car causing a crash or an automatic trading algorithm causing financial losses or decisions that are legally or morally wrong causing fines or brand damage.

In the following sections, we begin from the objective "automated decisions driving automated actions" and solve back to the means and technical approaches required to achieve that objective.

## Actions Require Decisions

Companies and organizations make a large number of decisions[3] day in and day out and take a large number of concrete actions based on these decisions.

Big strategic decisions are made by leaders and boards, for example, whether to acquire a company in order to expand a business, how to shape corporate culture and image, and how much overall risk to take in balance with revenue objectives. These strategic decisions may be informed by leveraging AI techniques; however, the final judgment and decision remains the responsibility of human beings – and this will not significantly change. Ultimately, leaders and boards remain responsible for these strategic decisions and all their consequences, but AI may help them make better decisions. There is usually not a high quantity of decisions of this kind, and they are well suited to be made by humans after due consideration and discussion.

However, a by far larger volume of decisions in an enterprise or organization – it may be millions or billions – need to be taken consistently, quickly, and with high frequency based on data, guidelines, and constraints. Examples of such frequent decisions are whether to open a door for a person who wants to enter, what to offer in a contextual

---

[3]See [4] for more information on the theoretical and application of decision analysis.

marketing campaign, optimization of agent-client interactions, deciding whether an autonomous car should break in in a given situation, deciding whether to approve or reject an insurance claim or loan request, and facilitating buying decisions.

The following are some typical examples of large-quantity and high-frequency decisions that we will explore more in the following sections:

- **Next best offer**: Deciding what products to offer to a customer when logging on to a website

- **Accurate travel information**: What flight departure and arrival times to display on airport screens and websites

- **Floor production optimization**: Whether to hold a production line on a factory floor

These kinds of decisions[4] are not optimally made by humans, because the quantity of these decisions it too large and the input parameters for these decisions are complex, making it infeasible to staff making these high-volume decisions with humans.

In order to automate these kinds of high-volume data-driven decisions, in the past, typically inflexible deterministic programs were used that required developers to formulate deterministic algorithms and pre-defined rules to process input parameters and determine the desired decision. This required developers to be available to change code when needed, even for small adjustments, which can be required rather often over time. Or humans would still determine these decisions following rules and guidelines defined for them and/or based on their own judgment, allowing adjustment to changing circumstances but requiring significant staffing causing high cost and needing more time per decision to be made.

## Decisions Require Predictions

In order to determine high-volume decisions with a degree of flexibility and yet in an automated fashion, data-driven predictions are needed. These predictions can be generated using state-of-the-art AI techniques, using relevant data to fuel a process that accesses and feeds data into a predictive model and obtains the resulting predictions, as you can see in Figure 1-1.

---

[4]See [5] for more information on automated decision making.

***Figure 1-1.*** *From Data to Predictions*

This predictive insight then serves as one of the basic inputs for the decision-making process. The following are just a few examples of predictive insight, corresponding to the three examples of required decisions we gave in the previous section:

- **Likely product interest to inform next best action**: What products customers likely are interested in and ready to buy next, and what is the likelihood of a customer buying one or even several of the proposed products. This prediction is needed to decide what product or service to offer next to the customer, guaranteeing the highest likelihood for acceptance.

- **Projected flight arrival at the gate to inform accurate travel information**: When a plane that is still in flight will arrive at the gate, given all circumstances, such as air traffic, head or tail winds, taxi time, and many more. This prediction is needed to decide what arrival time to display for an inbound flight.

- **Predicted quality of parts for floor production optimization**: Whether based on current sensor information, the next batch of parts produced by a machine in a plant will be good or faulty. This prediction is needed to decide whether to continue production or hold the production line.

However, predictions alone are typically not enough to make smart, ideally optimal decisions. For example, using a prediction of a customer's interest in a product just by itself could result in a problematic decision, such as offering something that is not in stock and then disappointing the customer with a long wait time. You need more than just predictions to make *smart* decisions.

# Smart Decisions: Prediction and Optimization

Often, decisions and actions cannot be taken purely based on individual predictions. Revisiting the previous example, what product really makes sense to offer to which customer, apart from the customer's interest, can also depend on stock in warehouses and time to deliver, profitability by product, size of marketing budget, customer status, acceptance or denial history of past offers, and many other parameters.

In situations where decisions need to be made more holistically, not only based on predictions, but in the context of business constraints and other factors, predictions alone cannot adequately solve the problem. Sets of predictions need to be combined with optimization based on these predictions and additional data, constraints, and objectives in order to determine the optimal combination of decisions and resulting actions.

Figure 1-2 illustrates this flow from relevant data via predictive insight toward optimized decisions. Predictive ML and DL modeling in combination with decision optimization (DO) enable this flow.

***Figure 1-2.*** *Flow from Data to Optimized Decisions*

The AI discipline of ML established approaches to make predictions based on *data*, without requiring deterministic code for each different problem. ML algorithms build mathematical *models* based on sample data, which is used to train these models. There are a wide range of ML model types for different problem domains, which are trained by ML algorithms. Three main types of ML algorithms are supervised learning, unsupervised learning, and reinforcement learning algorithms.

For more information on ML, DL, and DO concepts, see *Chapter 3, "Key ML, DL, and DO Concepts."*

# Data Fuels AI

In order to create and train *good* ML and DL models, we need good data as the very foundation of any AI-based solution. Data must be sufficiently relevant, accurate, and complete to be useful for training models. Also, data needs to be representative and reflect reality, avoiding undesired bias[5]. Only if qualitative and relevant data is used to train ML and DL models can accurate and precise models be created as a result of the training process. If data is not appropriate and sufficiently representative of a given business scenario, resulting models will typically perform poorly and/or be biased.

# Garbage In, Garbage Out

It may sound trivial, but if poor data is used to train ML and DL models, garbage in will cause garbage out. A good example is an anecdote from an early image recognition project, where data scientists used images with tanks and images without tanks to train an artificial neural network model to recognize images with tanks. The model was trained with a set of labeled training images and then validated with a set of labeled validation images. The resulting model seemed to work fine. Later, it was further tested with new set of images and then performed poorly. In real projects, this phenomenon does occur more often than not.

After some analysis, the data scientists found that regardless of tanks present while the initial set of images were taken, the labels were strongly correlated with sunny or cloudy weather conditions. As a result, the output of the artificial neural network (ANN)[6] model was strongly influenced by whether pictures were taken when it was sunny vs. cloudy. Only after repeating the training process with an improved and more adequate labeled training data set, the ANN model could achieve better results.

Generating a qualitative and representative set of labeled data sets for training as well as validation and testing is an essential and often time-consuming task for data scientists and data engineers.

---

[5]Depending on the business context and use case, bias in ML or DL models may be an anticipated aspect. However, in most scenarios, bias should be avoided.

[6]We provide a high-level description of artificial neural networks (ANNs) in Chapter 3, *"Key ML, DL, and DO Concepts."*

# Bias

If biased data is used to train a model, the resulting model will likely also demonstrate bias. For example, a bank may want to automate decisions in a loan approval process, by taking a sample of previous loan decisions made my humans and using these to train a model to automate these decisions going forward. If the data set of past human decisions is without bias, the model can be expected to be fair.

However, if the past human decisions in the training data set were biased toward more likely declining loans for some demographical characteristics (e.g., age, race, ethnicity, gender, marital status), the model would be trained to be biased as well.

Even if the entirety of past human decisions was not biased, if the training data set is sampled in a way that is not sufficiently representative, the resulting model could still have bias[7].

# Information Architecture for AI

Given how critical data is for AI, it is essential to establish a proper information architecture (IA)[8] for managing data that will be used to fuel AI and analytic assets created to build AI.

This involves trusted processes for *Data Ops*, *data science*, and *Model Ops* working hand in hand, as it is illustrated in Figure 1-3.

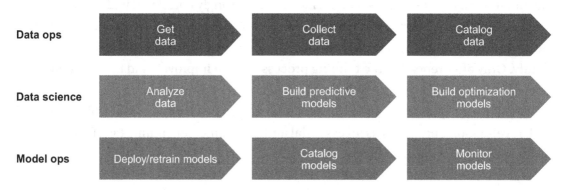

***Figure 1-3.*** *Data Ops, Data Science, and Model Ops*

---

[7]We elaborate on bias including how to monitor bias in Chapter 4, "*AI Information Architecture,*" and Chapter 6, "*The Operationalization of AI.*"

[8]We describe key concepts of an information architecture for AI in Chapter 4, "*AI Information Architecture.*"

The following is a short description of these three processes.

- **Data Ops** involves getting data from all relevant sources, collecting data in readily available and performant data stores, and cataloging data with a level of verification, making sure that the data is representative and not biased, if sampled that it is not skewed in the sampling process, and – last but not least – managing data to be reliable and immutable in order to support reproducibility. For any model trained based on data, a good information architecture should make sure the model can always be traced back not only to the code that was used to train it but also to the actual data that was used to train, validate, and test it and the lineage or provenance of that data. Even new data used for subsequent retraining processes, and additional adaptations of the model, needs to be taken into consideration.

- **Data science** is required to get from data to models. Data science skills and often subject matter expertise are needed to get from the data provided by the Data Ops elements of an information architecture to predictive and/or prescriptive models, which are made available for use by processes and applications through Model Ops. This typically involves analyzing data to really understand it enough to see whether it's the right data to use, possibly labeling of data which may require special subject matter expertise, building and validating predictive models, and possibly additionally building optimization models to work hand in hand with data and predictions.

- **Model Ops** involves deploying models for use in production and if needed retraining models, and monitoring models in production. Similar to dealing with data, the information architecture should also treat models and model deployments as first-class entities that need to be cataloged to be well-known entities in the system. Model Ops typically also needs to be connected to Data Ops, to feed new data to in-production models in a reliable and performant fashion, for example, for real-time scoring.

# Putting It Together: The AI Life Cycle

We have observed that in order to achieve the objective of informing and automating decisions and actions in an enterprise or organization, we need predictions and often also optimization. In order to make predictions, we need to train machine learning models which requires good, trusted data. In order to collect and manage a body of good, trusted data, we thus need an AI information architecture.

While we started our reasoning from the objective of enabling automated actions and solved back to the means to achieve it, in order to build working solutions, we need to start from data and drive from data to predictions to optimal actions.

This practical approach to achieve outcomes has been captured in the ML life cycle[9], as depicted in Figure 1-4. We further elaborate on various ML life cycle aspects in the context of the information architecture in Chapter 4, "*AI Information Architecture.*"

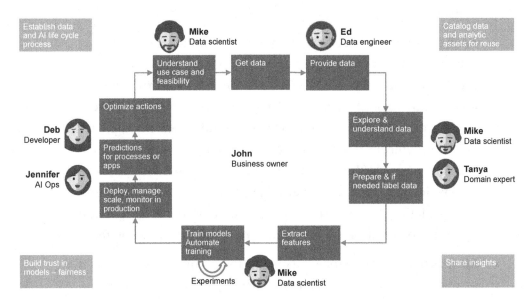

***Figure 1-4.*** *End-to-End Data Science/ML Cycle*

---

[9]See [6] for more information on the AI life cycle.

## Understanding Use Case and Feasibility

The first phase in the ML life cycle typically is to start by understanding the use case and whether it is in practice feasible to solve. Key aspects to define and clarify are what is the exact objective of a project, what are the concrete decisions and actions that are to be automated, and feasibility of solving the use case with state-of-the-art AI and ML approaches and with the data that can be made available and be used in practice.

None of this can be taken for granted: Many problems are not yet viable to solve with current ML approaches. Often it may not be possible or allowed by law or policies to actually get the data that would be needed, for example, due to data privacy reasons, or some required data may simply not yet be collected and stored.

Coming out of this phase, there should be a good degree of clarity on what problem to solve and what data to use to solve it and confirmation that the required data can indeed be obtained and consumed in practice, combined with required confidence that existing ML techniques can be applied.

## Collect Data

The next phase of the ML life cycle is collecting raw data to have a sound foundation to work from. Typically, in an enterprise or other organizations, there will be diverse data sources, some inside and some outside, some structured and some unstructured, and some already in qualitative and adequate shape and some needing refinement in order to be useful.

It is important in this phase to properly catalog data so that in subsequent steps it is easy to find, access, and use it in data science and ML projects. Depending on the source of the data, it may make sense to reference data or create copies of data or samples of data in a data warehouse or data lake and in either case catalog the data for further use.

Sometimes in this phase, it may be necessary to omit or anonymize sensitive data, for example, names, social security numbers, credit card numbers, and so on, before making the data available to data scientists for the following steps.

## Explore and Understand Data

The next phase is typically to begin to explore, analyze, and understand the available data, for example, to find relevant patterns and anomalies and to understand the statistical distribution of the data and its coverage over time, by geography, by demography, and many more.

In this phase, data should also be checked for bias, for example, if using historical credit approval decisions to train a credit approval model, already during this phase, it makes sense to check for bias, for example, based on gender or age, ZIP code, or other attributes that should not influence how the model works later on. Nevertheless, depending on the use case scenario, bias may actually be a desired characteristic. For instance, ML models for fraud discovery of debit card usage in ATM or POS networks may very well take into accounts a ZIP code or certain stores, where fraud may occur more often or in certain scenarios.

It is critical to really understand the data well and ensure its representative of a domain before venturing into the subsequent steps. Data science and ML projects can easily get stuck in later phases if the de facto available data and what can be done with it is not sufficiently understood.

Often, it may be necessary to repeat or expand the data collection process and collect more representative data, obtain additional permissions to source more data, or even refine the use case definition to make it solvable now having a better understanding of the available data.

# Prepare and If Needed Label Data

The next phase is to prepare and if needed label data, in order to get to data sets that can effectively be used to train models. Data labeling, which means to tag or annotate data samples, can be a very challenging and time-consuming effort.[10] In addition, it may be necessary to, for example, cleanse the data to remove invalid or erroneous data items and align data encoding if inconsistent to achieve a good level of data quality. Also, it may be necessary to sample data in a way that is representative and not biased.

For instance, if a raw data set in an HR use case is skewed in having too many data points about males and too few data points about females, this would likely result in biased models later on. In order to fix this problem, either more data points about females could be collected and added, or if there are plenty of data points, the number of data points about males could be reduced to be on par.

Often, for example, for supervised ML, labeled data is required. This may require involving subject matter experts in a project if labeling requires high skills, for example, labeling X-ray images with diagnostics or labeling loan application data with eligibility,

---

[10]See [7] and [8] for more information on data science related data labeling.

or in some other cases may just require normal human cognition, such as labeling cars or pedestrians in images. For the latter, it is possible to leverage third parties that offer labeling services. The accuracy and quality of labels for data is ultimately critical for the accuracy and quality of resulting models, which may require to have multiple outstanding subject experts label the same data and determine the right label based on the combined input.

## Extract Features

In this phase, data scientists extract features from data, that is, they determine which aspects of the input data are relevant for predictions or classification and which should be used as part of model training. This step may not just be purely data driven, based on existing laws, regulations, and guidelines within a company or organization, certain attributes of the input data may not be allowed or not be desired to influence ML or DL model decisions.

A credit approval model, for instance, should not discriminate based on gender; consequently, a gender attribute that may be included in the data should not become a feature to be used in training a model for credit approvals.

## Train and Validate Models

In order to train models, typically initially data scientists will explore many options and run a range of experiments to identify good candidate model pipelines, validate the resulting models, and determine model quality and performance metrics. This can be very compute and memory resource intensive, depending on data size and ML or DL algorithms being used. To facilitate model training by data science teams in an enterprise, typically public cloud or private cloud approaches are used, allowing to allocate compute and memory capacity from a pool of resources on demand.

After training and validating models, the most promising models should be tested thoroughly with more data, eventually leading to a model that the data science team is confident works well and complies with relevant laws and guidelines identified as relevant earlier in the project.

# Model Reviews and Approvals

For critical use cases, models typically need to go through a model review and approval process. For example, in banks, insurances, and healthcare organizations, a model could influence millions in revenue or risk, thousands of healthcare coverage decisions, or thousands of diagnostic assistance recommendations. To avoid costly or dangerous consequences of deploying new models or new versions of models, data scientists may need to provide extensive documentation of how a model works and based on what inputs it makes decisions, to be reviewed by other data scientists, business stakeholders, legal functions, and many others until all required sign offs are given for the model to proceed to deployment and operationalization for usage of the models by applications or business processes.

Model risk management solutions[11] can help streamlining the documentation and approval process and to account for and track model risk.

# Deploying and Monitoring Models in Production

Crossing the divide between data science and experimentation performed by data science teams on one side and rigid production deployment processes governed and run by IT departments on the other side is a challenge in many projects. In order to bridge from working environments in which data scientists experiment and create initial data and model, including model pipeline artifacts to tightly controlled test and production deployment systems, proper hand off is required from data scientists to operational IT staff.

In many cases, working environments used by data scientists and IT systems where models ultimately are deployed are completely separate systems. For example, the working environment for data scientists may be a cloud-based software-as-a-service solution or a software solution on a Kubernetes cluster in a private cloud. The production systems where models need to be deployed may be applications in cars, servers on a factory production floor, or a model scoring service in the context of an application or business process on a public or private cloud operated and run by a completely different organization.

---

[11]Please refer to Chapter 8, *"AI and Governance,"* for more information on risk management in the context of AI.

In order to connect across, often data scientists need to deliver their model training code (or model training flows if using a visual tool) and all required source artifacts into a code repository, for example, an enterprise Git service. This can enable an IT process where after any required sign offs a continuous integration/continuous delivery (CI/CD) pipeline can be set up to train production-ready models and model pipelines using the training code or flows from the repository, store these models and model pipelines in a model repository, and then from there deploy to a test system and ultimately to production systems.

The operationalization of AI models and artifacts into an existing IT and application landscape is often neglected during the development cycle. Since enterprises are indeed struggling with the deployment and operationalization, we have devoted a whole chapter[12] on this particular topic.

# Predictions for Applications or Processes

In order to use models in the context of applications or business processes, model deployment services need to match the quality of service of the consuming applications and processes. For production grade applications that need to operate with high availability and disaster recovery, the models used also need to be deployed with matching HA and DR characteristics. In order to achieve this, ML and DL models may need to be deployed on multiple independent systems in different sites or even in different regions, so that requests for model scoring from applications or business processes can be load balanced across deployments with failover if one deployment becomes unavailable. For large-scale use of nontrivial models, model scaling is needed in order to allow the model to run on as many cores with as much memory as needed.

Once a model is deployed to production, its operation needs to be constantly monitored to ensure the model deployment meets response time constraints and performs within expected parameters. In many use cases, models input and output also have to be monitored and analyzed continuously to detect any bias the model may have or drift a model may develop over time when input data over time differs too much from training data that was used to train the model.

If issues are detected, in certain use cases and situations, local retraining may be performed on the production system to adjust the model. Nevertheless, in many cases, as a result data scientists may have to be engaged again to create a new more robust model and then submit it to the production deployment process again.

---

[12]Refer to Chapter 6, "*The Operationalization of AI.*"

# Optimize Actions

As we observed earlier, often a prediction coming out of an ML or DL model alone is not enough to derive good actions. What decisions to make and what actions to take based on predictions usually depend on the context: other predictions, additional data, and constraints. In simple cases, some application or process logic can wrap the use of models and factor in context to determine decisions and resulting actions. However, this will usually not lead to optimal results.

As we have briefly mentioned at the beginning, in order to continuously optimize decisions in the context of a range of predictions, data, and constraints, decision optimization (DO) can be applied. For example, an application or business process can use deployed ML models to make batches of predictions. The application or process can then invoke solving of a DO model with these predictions combined with context data and constraints in order to determine the optimal set of decisions and then put these decisions into action.

Figure 1-5 is an illustration of this process. Chapter 5, *"From Data to Predictions to Optimal Actions,"* elaborates on this coherency between various data input streams, corresponding predictions, and optimized actions.

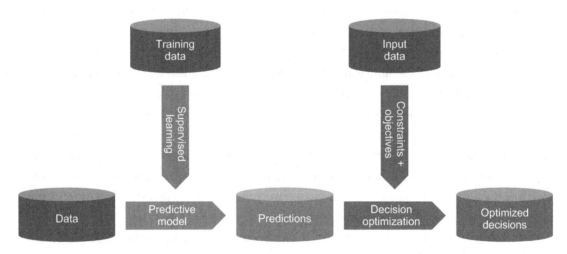

***Figure 1-5.*** *Toward Optimized Decisions*

# Reap the Benefits of Automated Actions

If you manage to establish and operate the end-to-end AI life cycle as outlined earlier for your enterprise or business area, unprecedented business acceleration and efficiency can be achieved. Decisions that used to take time can be made immediately if fully automated or significantly faster when human decisions are augmented by AI, to drive to automated or AI-augmented actions. As a result, business process durations can be shortened, and output and customer satisfaction can be increased.

# AI and Cognitive Computing

AI and cognitive computing (CC) are very much related to each other. Artificial intelligence is intelligence demonstrated by machines that mimic *cognitive* functions that humans associate with other human minds, such as *learning, reasoning,* and *problem solving.* Thus, AI is composed of many domains, such as ML and DL, DO and rule-based systems, and so forth to deliver, for instance, predictive and prescriptive insight.

Cognitive computing deals with systems where the emphasis is on the ability to learn behavior and decision making through experience, including to learn from scratch or based on feedback, and even education or training. CC focuses on the cognitive capabilities and supports multiple forms of expression that are more natural for human interaction, where the primary value of CC is learned expertise. With CC, the emphasis is to continuously learn and evolve as new experience, scenarios, and responses become available.

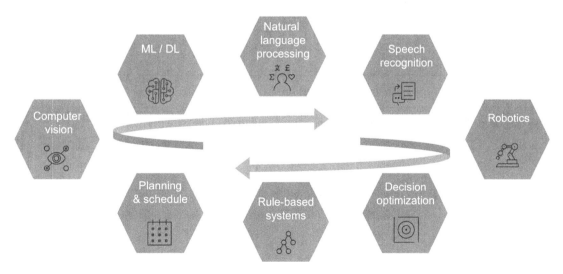

*Figure 1-6. AI and CC Domains*

In a way, AI and CC are two sides of the same coin. Figure 1-6 depicts the various domains that make up AI and CC. As you can see, the AI and CC scope is rather broad. In this book, we touch on most of these domains, but our focus is more on the enterprise relevance. For instance, we do not intend to elaborate on *robotics* and *computer vision*.

In Chapter 13, *"Limitations of AI,"* we will especially detail out today's limitations of AI in terms of its cognitive capabilities.

# AI, Blockchain, Quantum Computing

AI is not the only domain providing enterprises with business benefit. Blockchain is making more than headway, establishing itself as an essential immutable, shared ledger application which can be used to record transactions of various assets. In Chapter 11, *"AI and Blockchain,"* we share with you our view on the intersection of AI with blockchain. Although a reality in research and development, quantum computing appears a bit further on the technology horizon – at least as far as its deployment and usage in today's enterprises is concerned. Nevertheless, in Chapter 12, *"AI and Quantum Computing,"* we provide you with our thoughts regarding some AI problems, which are likely to benefit from quantum computing.

# Key Takeaways

We conclude this chapter with a few key takeaways, which are summarized in Table 1-1.

***Table 1-1.*** *Key Takeaways*

| # | Key Takeaway | High-Level Description |
|---|---|---|
| 1 | AI objective: Automate actions | Automated actions require automated decisions. When based on good decisions, outcomes can be great, but if decisions are not good, significant damage can result |
| 2 | Good decisions are not trivial to make | Predictions can help make automated decisions, but are not enough for good decisions. Decisions need to be made considering other decisions and context |

(continued)

***Table 1-1.***  (*continued*)

| # | Key Takeaway | High-Level Description |
|---|---|---|
| 3 | Smart decisions with predictions + optimization | Optimizing decisions based on data + predictions + constraints allows to make smart decisions that make sense together |
| 4 | Machine learning to make predictions | Machine learning can make predictions based on data, without requiring specific code for each different problem |
| 5 | Data fuels AI – garbage in, garbage out | Data is critical for AI and especially for machine learning, requiring a solid information architecture to ensure reliable quality data |
| 6 | Information architecture | An information architecture for AI involves Data Ops connected with data science and Model Ops processes |
| 7 | Putting it together: AI life cycle | The AI life cycle involves connecting and collecting data, understanding and refining data, building predictive and prescriptive models using data, and deploying, running, and monitoring models in production |

# References

[1]   Pandya, J. Forbes. *What Is The Future Of Enterprise AI?* 2019, www.forbes.com/sites/cognitiveworld/2019/11/17/what-is-the-future-of-enterprise-ai/#1fea7eda7a79 (accessed April 28, 2020).

[2]   Elliot, B., Andrews, W. Gartner. *A Framework for Applying AI in the Enterprise,* www.gartner.com/en/doc/3751363-a-framework-for-applying-ai-in-the-enterprise (accessed April 28, 2020).

[3]   Earley, S. *The Ai-Powered Enterprise: Harness the Power of Ontologies to Make Your Business Smarter, Faster, and More Profitable.* ISBN-13: 978-1-928055-50-1, Ingram Publishing Services, 2020.

[4]   Ishizaka, A., Nemery, P. *Multi-criteria Decision Analysis: Methods and Software.* ISBN-13: 978-1119974079, Wiley, 2013.

[5]   Delen, D. *Prescriptive Analytics: The Final Frontier for Evidence-Based Management and Optimal Decision Making.* ISBN-13: 978-0134387055, Pearson FT Press, 2019.

[6]   Joakar, A. Data Science Central. *Explaining AI from a Life cycle of data,* www.datasciencecentral.com/profiles/blogs/explaining-ai-from-a-life-cycle-of-data (accessed April 28, 2020).

[7]   The Economist. *Data-labelling startups want to help improve corporate AI.* www.economist.com/business/2019/10/17/data-labelling-startups-want-to-help-improve-corporate-ai (accessed April 28, 2020).

[8]   Medium. Data Annotation: *The Billion Dollar Business Behind AI Breakthroughs.* https://medium.com/syncedreview/data-annotation-the-billion-dollar-business-behind-ai-breakthroughs-d929b0a50d23 (accessed April 2020).

# CHAPTER 2

# AI Historical Perspective

Without us being fully aware of and constantly appreciating it, AI is already impacting us since years, even decades. Therefore, *an AI historical perspective* doesn't seem to be a vital consideration any more: AI has established itself as an undeniable fact of life. Its impact is already noticeable to each and every individual and the society as a whole.

So what has changed, and why do we still intend to provide an AI historical perspective? The transformation of AI from a primarily academic and research domain toward a broad set of commercially relevant AI applications, the availability of thousands of freely downloadable open source libraries, and advancements in specialized processors (e.g., GPUs, FPGAs, ASICs) represent some significant changes that occurred in the last decade.

Even today, however, there are scenarios where there is simply no need for AI because the problems or issues are too simple or that the problems are too sophisticated for today's AI capabilities. These aspects will be discussed in Chapter 13, *"Limitations of AI."* In addition, there is often a certain hesitation to use AI (ethical concerns, regulatory requirements, etc.); these aspects are discussed in Chapter 8, *"AI and Governance."*

In this relatively short chapter, we intend to provide a brief historical perspective, a glimpse of the recent past, where AI capabilities were still somewhat immature and consequently the AI applications were relatively limited. This *historical perspective* should therefore shed some light at this gap between the emergence of AI in the 1950s and 1960s and the explosion of AI applications in recent years and why it has taken so long – half a century – to adopt AI.

© Eberhard Hechler, Martin Oberhofer, Thomas Schaeck 2020
E. Hechler et al., *Deploying AI in the Enterprise*, https://doi.org/10.1007/978-1-4842-6206-1_2

# Introduction

Advancements and adoption of technology usually happen gradually; it is an evolutionary approach. It may therefore be challenging to determine a date when there was no AI. The term AI was coined by John McCarthy[1] in 1955; however, concepts, thoughts, ideas, and even algorithms that led to what later became known as AI existed much earlier[2]. Here are just a few examples: Bayes' theorem on conditional probabilities goes back to the British mathematician Thomas Bayes, who lived in the eighteenth century. The least square method has been developed more than 200 years ago; the stochastic concept of Markov chains has been developed at the beginning of the twentieth century. Even the first artificial neural networks (ANNs) have been implemented at the beginning of the 1950s, based on concepts from the 1940s.

These early AI developments and advancements were more of theoretical nature, rather than mature and proven. Consequently, they were less applicable in the field and remained – more or less – within an academical circle.

In this chapter, we intend to discuss this gap and illustrate why the theoretical AI underpinning of the second half of the twentieth century led to the remarkable AI boost in the last decade.

# Historical Perspective

Although AI including key ML and DL concepts is dating back several decades, in the 1960s and 1970s, there were limited AI capabilities available that had noticeable relevance in the field. In this section, we discuss how technological AI advancements have changed the way AI can be used and what impact this had on business, individuals, and the society as a whole.

We also venture a glimpse into the future by briefly elaborating on *beyond AI*, meaning what may change in the near future. This is further taken up again in Chapter 14, *"In Summary and Onward."*

---

[1]See [1] for more information on the Dartmouth summer research project where the term AI was created.

[2]See [2] for a very short history on AI.

# Technological Advancements

As it is true with all domains, it applies to AI as well that there are novel approaches and methods and innovative technological advancements, for instance, improvements in HW processor speed and memory sizes, emerging specialized processors (e.g., GPUs, FPGAs, ASICs, etc.), which – over time – enable AI applications in a much richer and more relevant way. This *historical perspective* is therefore an attempt to look at AI from an evolutionary and motivational perspective: to understand what were the AI capabilities and application areas a few years back or some decades ago (given the technological status quo at that time) vs. what is possible today (with recent AI and technological advancements) and what might even be ahead of us in the near future (based on promising research and development efforts).

Positioning ourselves to AI has always been more or less determined by the state-of-the-art AI technology and the applicability of AI in the field. In the first decades after the term AI was coined, and pretty much throughout the second half of the twentieth century, AI and its siblings were mostly associated to academic and research activities. As a result, organizations and individuals were considerably less impacted by AI; the commercialization of AI and visibility of AI in the field were rather limited.

During the last decade of the twentieth century, we already saw major AI advancements, such as IBM's Deep Blue[3] system, who beat the world chess champion after a six-game match. However, substantial increases of AI enhancements with sufficient maturity and readiness for deployments in the field were visible in the twenty-first century, particularly in the last decade.

Table 2-1 lists the major milestones of AI focusing on the last decade. As you can see from this (incomplete) list, organizations and individuals became increasingly impacted by AI; the commercialization of AI is well on its way, and *not to AI* is clearly no option anymore.

---

[3]See [3] for more information on IBM's Deep Blue system.

***Table 2-1.*** *Major Milestones of AI in the Last Decade*

| # | Milestone | Description | Impact |
|---|-----------|-------------|--------|
| 1 | IBM Watson[4] | IBM Watson played Jeopardy! and won against human opponents | This clearly demonstrated that AI is better than humans at natural language-related games like Jeopardy |
| 2 | AlphaGo Zero[5] | Combination of advanced search tree with ANNs won against Go champions | First success that ANNs can learn games like Go from scratch, without initial training with labeled data |
| 3 | Face recognition[6] | Success in training a face detection ANN without having to label images (unsupervised learning) | First success with artificial neural networks (ANNs) that can learn human face recognition without being explicitly trained using labeled image data |
| 4 | Voice assistants[7] | Apple Siri, Amazon Alexa, Google, and others are 2nd-generation voice assistants | Transformation toward voice-based computer interaction (voice-controlled AI), where assistants are able to understand and respond to statements and questions in natural language, after performing a set of actions |
| 5 | Tesla[8] self- driving vehicle | Tesla announcement that cars produced after November 2016 have full self-driving HW | Far-reaching impact for the automobile industry, drivers, and the transportation industry |

[4]See [4] for more information on IBM Watson's win on Jeopardy!

[5]See [5] for more information on AlphaGo and AlphaGo Zero.

[6]See [6] for more information on the face detection problem.

[7]See [7] for more information on voice assistants.

[8]See [8] for more information regarding Tesla's announcement on self-driving hardware.

These major milestones are depicted in Figure 2-1.

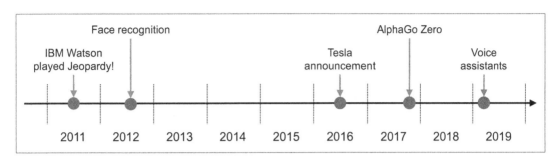

***Figure 2-1.*** *Major Milestones of AI in the Last Decade*

Further AI enhancements were made available by the open source community with open source software licensing[9] and by vendors, such as Apple, IBM, Facebook, Google, and others, which has literally led to an explosion of AI applications, especially in the last decade.

# The Evolution of AI

Until now, we have elaborated on the emergence of AI in the 1950s and 1960s, including the latest key milestones of AI in the last decade. This evolution of AI is characterized by an ever-increasing commercialization and presence of AI in the field, impacting practically all walks of life. It is clearly a transformation from AI being primarily restricted to an academic and research domain toward an explosion of mature and usable AI applications. During the past 50 years, the limitation to apply AI in practical terms has obviously changed into a rapidly expanding and commercialized usage of AI.

The evolution and commercialization of AI continues at an unprecedented speed, with major AI deficiencies being addressed by research and vendors alike. Reinforcement learning without initial training with labeled data, generalization and multitask learning, autonomous AI, and – most importantly – enriching AI with cognitive capabilities, such as reasoning and understanding, represent just a few examples of where AI will evolve toward in the future. By then, we see AI embedded in an all-embracing fashion without us being aware of its presence.

---

[9]See [9] for more information on open source software licensing.

In Chapter 13, "*Limitations of AI*," we further elaborate on the current AI limitations and describe key research efforts in an attempt to overcome some of those limitations. In Chapter 14, "*In Summary and Onward*," we provide an outlook into a determined AI-pervasive world.

# Some Industry Examples

In regard to AI, most enterprises and individuals alike have already begun their journey to AI. However, there are differences. In a recent survey by O'Reilly[10], for instance, 49% of respondents indicated the stage of the ML adoption in their organization as *exploring*, 36% as *early adapters*, and only 15% as *sophisticated*.

Regardless of how this diverse AI adoption picture may improve in the future, today it holds true when we look at various industries: although AI is present in all industries, the decision to move to AI, or rather not, differs somewhat by industry. This is mainly related to the different opportunities, characteristics, and business models of the industries. We focus on the following six industry sectors by proving an (incomplete) exemplary description of the AI usage and highlighting how AI has transformed these industries in recent years and – most importantly – what are the key AI trends in those industries:

- **Financial services industry**: In the financial services sector, fraud discovery and prevention is still mainly done via rule-based models, with ML models just about finding their entry into leading banks. AI is increasingly infused in the credit scoring and credit decision process. AI-infused risk management and AI-based regulatory compliance further increase compliancy and also reduce latency of insight. AI further enables customized and targeted customer interaction to improve the personalized banking experience. Security for banking transactions and customer interaction will become greatly improved by AI. We will see major advances of AI usage to discover and prevent fraudulent activities, such as identity theft, investment fraud, credit and debit card fraud, money laundry, fraudulent disbursements, and many more.

---

[10]See [10] for more information on the survey conducted by O'Reilly.

- **Automotive industry**: In the automotive industry, the adoption
  is mixed, as some automobile manufacturers are spearheading
  progressive usage of AI for full-fledged autonomous driving, while
  others are more hesitant and look for low-risk entry scenarios, such
  as using intelligent parking assist or traffic jam assist systems or other
  advanced driver assist systems[11]. Although autonomous driving AI for
  further optimization and autonomous manufacturing will represent
  a major shift in the automotive industry, there are additional
  disruptions, which represent a paradigm shift in how we think about
  transportation, such as self-driving transportation passenger services
  or AI-infused goods delivery services. AI and autonomous driving
  vehicles will be exploited as taxis, for commercial and personal
  transportation.

- **Aerospace industry**: In the aerospace industry, which is a relatively
  conservative industry when it comes to applying new technologies in
  general and particularly in the cockpit, adoption of AI may be more
  advanced in regard to customer service and ticketing, marketing,
  and passenger identification (facial recognition), rather than using
  AI on the flight deck and within the avionic control systems. Even
  predictive MRO (maintenance, repair, and operating) supplies are
  still an emerging domain for aviation[12]. AI still has to find its way into
  the flight deck, complementing autopilot technology and providing
  intelligent assistance for pilots.

- **Healthcare industry**: AI has already become established in the
  healthcare industry. AI-powered predictive analytics and pattern
  recognition can support the clinical decision process and improve
  accuracy of identifying patients being at risk of developing a certain
  cluster of symptoms. Robots are used in healthcare in the last
  three decades to assist surgeons. IBM Watson Health, for instance,

---

[11]See [11] and [12] for more information on advanced driver assist systems.
[12]See [13] for more information on AI in the aerospace industry.

advances healthcare with enterprise imaging and interoperability solutions. It also assists oncologists by assessing patient medical records in the context of *information from relevant guidelines, best practices, and medical journals and textbooks*.[13] The future of AI in healthcare may be seen by some related to the question when AI will not only assist but replace a human physician. However, the near-term future of AI in healthcare is more related to analyzing and discovering patterns and correlations in the vast amount of medical data, including information obtained from DNA; medical images, like CT scans, X-ray, and MRIs; and so forth to predict malignant tumors, vascular diseases, and other issues significantly earlier than it can be done today.

- **Manufacturing industry**: AI has long made its way into the manufacturing industry. Some of the more recent trends are AI-powered visual inspections based on relatively short learning cycles based on only a few product defect images. Optimizing and shortening the programming cycle of robots via AI-enabled automatically generated programs is another trend. While robots in manufacturing are still primarily programmed, future robots are able to recognize patterns, learn from past experiences, and understand visual and text input. Today's robots in manufacturing are programmed to perform specific tasks (they are actually much less intelligent as they may look); future robots can be trained and can learn – they are adjustable, and can truly collaborate with other robots, and take new instructions from humans that may complement the original programming of the robot.

- **Retail industry**: AI and ML are already loyal, long-time companions of the retail industry. They help to understand customer journeys through the shopping world, predict consumer purchasing patterns, and personalize consumer interaction and marketing. Some of the emerging AI trends in the retail industry are to provide very personalized product recommendations based on chatbot conversations or to better understand and discover patterns of behavior, poses, and movement through a shop to detect and

---

[13]See [14] for more information on IBM Watson Health and IBM Watson for Oncology.

prevent theft, especially for future unmanned stores. Innovative ways to deliver goods to consumers – for instance, via drones – are yet another AI-powered application that will significantly change the retail industry[14].

There are numerous cross-industry areas, where the hesitation to leverage AI, which was still prevalent two decades ago, has been shifted toward a clear application of AI: contextual targeted marketing campaigns, automation of general business areas, HR planning, governance, and risk and compliance – just to name a few.

# Key Takeaways

We conclude this chapter with a few key takeaways, which are summarized in Table 2-2.

***Table 2-2.*** *Key Takeaways*

| # | Key Takeaway | High-level Description |
|---|---|---|
| 1 | Technological advancements | The last decade has seen major technological AI advancements that have led to a significant increase and commercialization of AI applications |
| 2 | Origin of the AI term | The term AI was coined by John McCarthy in 1955; however, concepts and algorithms that led to what became known as AI were developed much earlier |
| 3 | AI evolution | Reinforcement learning without initial training with labeled data, generalization and multitask learning, autonomous AI, and – most importantly – enriching AI with cognitive capabilities, such as reasoning and understanding, are some of the hot evolving AI topics |
| 4 | To AI or not to AI | In the 1960s and 1970s, and throughout the second half of the twentieth century, AI was primarily an academia and research domain. Regarding commercialization, in the last two decades, the hesitation to apply AI has changed to a *how and when* to enter AI into use cases and applications |
| 5 | AI in various industries | Today, AI has been well established in some industries who are at the leading edge, while further AI advancements will increase the seamless exploitation and integration of AI in most if not all other industries |

---

[14]See [15] for more information on Amazon's Prime Air Delivery.

# References

[1]   McCarthy, J., Minsky, M.L., Rochester, N., Shannon, C.E. *A Proposal For The Dartmouth Summer Research Project on Artificial Intelligence*, 1955, `http://jmc.stanford.edu/articles/dartmouth/dartmouth.pdf` (accessed November 27, 2019).

[2]   Stanford University. *One Hundred Year Study on Artificial Intelligence (AI100)*, `https://ai100.stanford.edu/` (accessed November 27, 2019).

[3]   IBM. *Deep Blue. Overview*, `www.ibm.com/ibm/history/ibm100/us/en/icons/deepblue/` (accessed November 29, 2019).

[4]   IBM. *AI for the Enterprise. Why it matters that AI is better than humans at games like Jeopardy*, `www.ibm.com/blogs/watson/2017/06/why-it-matters-that-ai-is-better-than-humans-at-their-own-games/` (accessed November 29, 2019).

[5]   Silver, D., Hassabis, D. *DeepMind. Research Blog Post. AlphaGo Zero: Starting from scratch*, `https://deepmind.com/blog/article/alphago-zero-starting-scratch` (accessed November 29, 2019).

[6]   Ng, A.Y. et al. *Building High-level Features Using Large Scale Unsupervised Learning*, `https://icml.cc/2012/papers/73.pdf` (accessed November 29, 2019).

[7]   Vlahos, J. *Talk to Me: Amazon, Google, Apple and the Race for Voice-Controlled AI*. ISBN-13: 978-1847948069, Random House Books, 2019.

[8]   Tesla. *All Tesla Cars Being Produced Now Have Full Self-Driving Hardware*, `www.tesla.com/blog/all-tesla-cars-being-produced-now-have-full-self-driving-hardware` (accessed November 29, 2019).

[9]   Meeker, H. *Open (Source) for Business: A Practical Guide to Open Source Software Licensing – Second Edition*. ISBN-13: 978-1544737645, CreateSpace Independent Publishing Platform, 2017.

[10]    Lorica, B., Nathan, P. *The State of Machine Learning Adoption in the Enterprise.* O'Reilly Media, 2018.

[11]    Volkswagen. *Driver assistance,* www.volkswagenag.com/en/group/research/driver-assistance.html# (accessed November 29, 2019).

[12]    Bosch. *Invented for Life. Traffic jam assist,* www.bosch-mobility-solutions.com/en/products-and-services/passenger-cars-and-light-commercial-vehicles/automated-driving/traffic-jam-assist/ (accessed November 29, 2019).

[13]    Bellamy III, W. *Avionics International. Airlines are Increasingly Connecting Artificial Intelligence to Their MRO Strategies,* http://interactive.aviationtoday.com/avionicsmagazine/june-2019/airlines-are-increasingly-connecting-artificial-intelligence-to-their-mro-strategies/ (accessed November 30, 2019).

[14]    IBM. *IBM Watson Health products,* www.ibm.com/watson-health/products accessed November 30, 2019).

[15]    Amazon. *First Prime Air Delivery,* www.amazon.com/Amazon-Prime-Air/b?ie=UTF8&node=8037720011 (accessed December 2, 2019).

# CHAPTER 3

# Key ML, DL, and DO Concepts

Following the AI evolution in the previous chapter, this chapter is devoted to key concepts of machine learning (ML), deep learning (DL), and decision optimization (DO). We don't go into the details on the 101 of these concepts or mathematical and statistical science behind these areas; instead, we are discussing considerations about their practical application in enterprises or other organizations. It should serve as a high-level introduction for readers with limited knowledge in this space.

## Machine Learning (ML)

The AI discipline of machine learning (ML) established approaches to make predictions based on *data*, without requiring deterministic code for each different problem. ML algorithms build mathematical *models* based on sample data, which is used to train these models. *Machine learning (ML) is the study of computer algorithms that improve automatically through experience. It is seen as a subset of artificial intelligence. Machine learning algorithms build a mathematical model based on sample data, known as "training data," in order to make predictions or decisions without being explicitly programmed to do so.*[1]

---

[1]See [1] for a short overview description of ML and [2] and [3] for a more comprehensive treatment of ML.

© Eberhard Hechler, Martin Oberhofer, Thomas Schaeck 2020
E. Hechler et al., *Deploying AI in the Enterprise*, https://doi.org/10.1007/978-1-4842-6206-1_3

The most significant statement in this definition is *without being explicitly programmed*. In contrast to classic programming, ML enables computers to generate and train ML models based on training data provided to the ML algorithm. Sometimes, the terms *ML algorithm, ML model, training,* and *training data* can be somewhat confusing. We therefore provide you with a simple description on how these terms interrelate to each other: training an ML algorithm with training data generates an ML model. In other words, an ML model is the result when you train using an ML algorithm with training data.

Figure 3-1 is a simple illustration of this relation. This view is especially true for predictive ML models, for example, regression models, where the training data sets are labeled data records. Training means to minimize the error of the ML model using the labeled training data set, while avoiding overfitting in the process. Overfitting is when an ML algorithm produces a model that is overly adapted to the training data set. This would result in very high accuracy relative to the training data set, but typically lead to poor results when using the model with new data. In essence, after the training process, the ML algorithm yields an ML model that can be applied to new data to predict an outcome, as long as overfitting was avoided.

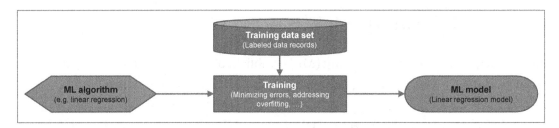

***Figure 3-1.*** *Training ML Algorithm with Labeled Data Set*

However, for some data preparation tasks, we are not necessarily generating an ML model. For instance, in order to reduce the dimension – and therefore the complexity – of the data space, a dimensionality reduction algorithm, called *principal component analysis* (PCA), can be used. For this scenario, the PCA algorithm is applied to an input data space, which yields an output data space with reduced complexity and dimensions. Another example is a k-means clustering algorithm, which is applied to an input data set (e.g., customer transactional records) to produce k groups or clusters of customers, for example, diamond, gold, and silver customer clusters. For these scenarios, there is no labeled data available and required. Figure 3-2 is a depiction of this concept.

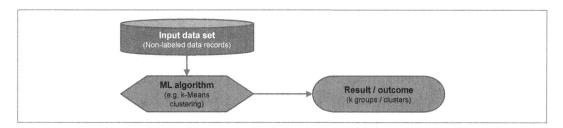

***Figure 3-2.*** *Applying ML Algorithm to Non-labeled Input Data Set*

As you have seen previously, there are distinct types of ML (or categories of ML) that address different problems or business domains. Figure 3-1 is characterizing what we call supervised learning, whereas Figure 3-2 is describing an example of unsupervised learning. In addition to the various types of ML, there are also different ML algorithms.

In the remaining part of this section, we introduce you to the key types of ML and types of ML algorithms.

# Types of ML

The following are the three main ML types: *supervised learning, unsupervised learning,* and *reinforcement learning.*

- **Supervised learning** focuses on training models that have a set of input variables (predictors) and a target variable to be predicted, where the training process continues until the model achieves a certain level of accuracy and precision in predicting the target variable on the training data. This requires labeled training data sets or training data sets that happen to already have the target attribute populated otherwise. Regression and classification are the key areas of supervised learning. Examples of predictions that are enabled by supervised learning are predicting what customers are likely to buy, when a machine will likely fail, classifying images or video based on content, and so on. In order for supervised learning algorithms to work, the training data needs to include desired output.

- **Unsupervised learning** does not involve any target variable and does not require a labeled training data set; rather its purpose is to cluster data points into different groups. Clustering and also dimensionality reduction (e.g., for feature reduction, structure discovery, etc.) are the key areas of unsupervised learning.

- **Reinforcement learning (RL)**[2] is based on learning via experience from trial and error. Reinforcement learning is an agent interacting with the environment to learn from a policy to perform certain tasks, further improving and optimizing its actions over time based on the rewards and punishments it receives. The agent learns to improve its actions based on the rewards it receives. RL is geared toward optimizing the learned policy. Learning an agent can be achieved through a variety of RL algorithms, such as model-free RL, for example, using *Q-learning*, or model-based RL, for example, using a *Markov decision process* (MDP).[3]

# Types of ML Algorithms

There exist quite a large number of ML algorithms, which can be applied depending on the use case to be implemented or data science tasks to be performed. These ML algorithms can be grouped into the following types of ML algorithms:

- Regression and classification

- Clustering

- Decision trees

- Bayesian

- Dimensionality reduction

- Artificial neural networks (ANNs)

- Ensemble

- Regularization

- Rule system

- Instance based

---

[2]See [4] for an introduction into reinforcement learning.
[3]See [5] for a short overview of the various RL algorithms.

We do not intend to describe all of these types of ML algorithms; however, we introduce you to a small subset of the most relevant ones in the context of this book. ANNs are covered in the section on DL further in the following.

## Regression and Classification

Regression models are used when a discrete value is being predicted. Business examples are stock price prediction, batch job elapsed time prediction, weather temperature prediction, or life expectation prediction in the context of a medical diagnostic analysis.

*Linear regression* is the most common form of regression analysis, where a line is drawn to best fit the data according to a mathematical criterion such as ordinary least squares. Other forms of regression analysis are *logistic regression, polynomial regression, stepwise regression*, and more.

Classification algorithms[4] are being used to predict a category, which can either be a two-class (binary) or a multi-class category. Business examples are fraud detection (fraud vs. non-fraud), spam email detection (spam vs. non-spam), churn prediction (churn vs. no churn), or customer classification (diamond vs. gold vs. silver). Additional examples are classification of drugs (medical), objects (self-driving vehicles), loan applications, and others.

There are quite a number of ML algorithms, which can be used for classification applications, such as *logistic regression, k-nearest neighbor, decision trees, support vector machines* (SVMs), etc.

*Support vector machines* (SVMs) are trained using examples marked as belonging to one of two categories, so that resulting models can predict whether a new example falls into one category or the other.

## Decision Trees

*Decision trees* can serve as predictive models where observations about an item represented in the branches lead to conclusions about the target values represented in the leaves. Decision trees can be used for regression and classification tasks.

As mentioned earlier, classification trees are used when the output is a discrete class, and regression trees are used when the predicted outcome is considered a real number. Ensemble methods such as *boosted trees* construct more than one decision tree. Algorithms for constructing decision trees usually work top down, choosing a variable at each step that best splits the items.

---

[4]See [6] for a short introduction of classification algorithms.

*Random forest* is an example of decision trees, which is an ensemble training method that can be used for classification, regression, or other tasks, where during training many decision trees are constructed and outputs are combined to determine the output class or value. Random forests can help improve relative to single decision trees when the model overfits to the training data set.

## Clustering

The goal of clustering algorithms is to group data into clusters to better organize the data. An important aspect of clustering is that the available data points are not labeled. Therefore, clustering belongs to the unsupervised type of ML. Evaluating the outcome regarding its correctness is subject to the business context, which may result into repetitively performing the clustering algorithm(s) till a satisfactory business outcome (set of clusters) has been found. Business examples are clustering customers (customer segmentation) based on their buying behavior, credit card usage pattern, travel-related preferences, and so on.

There exist quite a number of clustering algorithms[5], such as *k-means*, k- *medians*, *hierarchical clustering*, and *expectation maximization.*

## Bayesian

A *bayesian network* is an algorithm, which represents a set of variables and their conditional dependencies via a directed acyclic graph. Bayesian networks are ideal for predicting the cause of an event that occurred, for example, predicting a disease based on observed symptoms. Additional examples of bayesian algorithms are *naïve bayes classifier, gaussian naïve bayes, bayesian belief network* (BNN), and so on.

## Dimensionality Reduction

AI and machine learning are not only about developing AI models or clustering data. Prior to being able to develop AI artifacts of any sort, the available source data needs to be understood; the complexity of the data needs to be simplified; features need to be selected and transformed by, for instance, omitting features or predictors that do not contribute to the reduction of errors; and the dimension of the feature space may have to be reduced because of existing correlations among certain features. Dimensionality

---

[5]See [7] for a short introduction on clustering algorithms.

reduction is concerned about the selection and representation of the features to optimize and simplify the development of a classifier.

Because of its importance in data science, dimensionality reduction[6] remains a hot research area. Today, there are quite a number of algorithms available, such as *linear discriminant analysis* (LDA), *partial least squares regression* (PLSR), *mixture discriminant analysis* (MDA), *flexible discriminant analysis* (FDA), and *principal component analysis* (PCA), just to name a few.

PCA,[7] for instance, is an algorithm – often called a technique or method – used to reduce a large set of possibly correlated components or dimensions (predictors, features) to a set of uncorrelated (or less correlated) dimensions, possibly but not necessarily fewer dimensions, called principal components. Those dimensions are then orthogonal to each other (linearly independent) and ranked according to the variance of data along those dimensions. The goal is to extract the most relevant information from the data points along a set of new principal components to easily depict what accounts for the variation in the data.

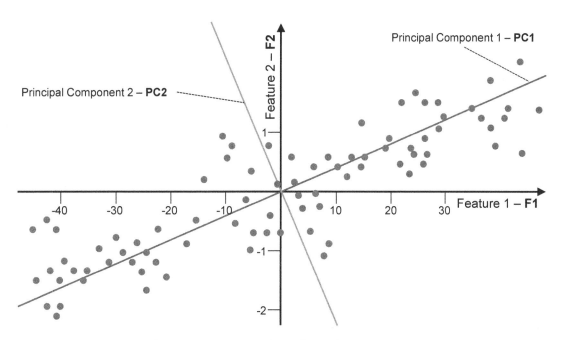

***Figure 3-3.*** *Principal Component Analysis (PCA)*

---

[6]See [8] for a short overview on dimensionality reduction.

[7]See [9] for a comprehensive and theoretical treatment of principal component analysis (PCA), including the mathematical background.

As you see in Figure 3-3, the two principal components PC1 and PC2 are the new dimensions, which are orthogonal to each other. They are derived as a linear combination of the original features F1 and F2. PC1 and PC2 are determined in such a way that the variances $var(PC1)$ and $var(PC2)$ – measure of the deviation from the mean for data points along PC1 and PC2, respectively – along the new dimensions are maximized with $var(PC1) > var(PC2)$. The covariance $cov(PC1, PC2)$ is taken into the equation to understand the relationship between values along the new dimensions.

The following is another example, where the original three-dimensional feature space can be reduced to a two-dimensional principal component space: imagine a set of data points distributed in a three-dimensional space, where the data points are approximately locatable on a two-dimensional plain. If you apply PCA to create two principal components PC1 and PC2 (of course, orthogonal to each other) that represent this plain, a third principal component PC3 may have such a low variance $var(PC3)$ that it can simply be neglected. Subsequently, you have reduced the original three-dimensional space to a two-dimensional space, which can now simplify the AI model development based on this "new" feature space.

As a result of the PCA algorithm, the complexity and number of correlated features may be reduced. However, this may come with the disadvantage of less interpretability of a subsequently developed AI model.

# Auto AI

A relatively new concept in machine learning is Auto AI or Auto ML, with the objective to automate the creation of machine learning pipelines and to automatically rank the resulting models to assist data scientists in finding good models more quickly.

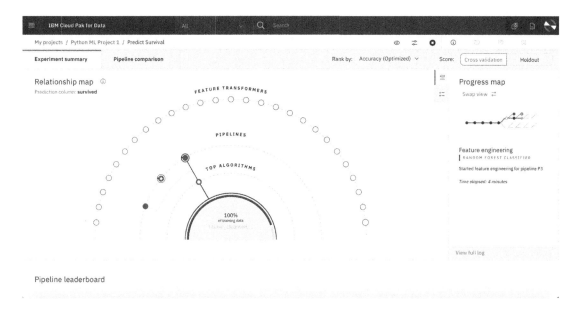

Pipeline leaderboard

***Figure 3-4.*** *Auto AI Training a Set of Models Using Multiple Training Pipelines*

In principle, users without data science skills may also use Auto AI to build models. However, ultimately expert judgment based on data science and subject matter expert skills is needed in order to assess whether an automatically created model is meeting the requirements of a particular use case and which in practice is really the most suitable model to use in a given business context.

An example of Auto AI is shown in Figure 3-4, based on the Auto AI function in IBM Watson Studio, which is running on IBM Cloud Pak for Data. The diagram shows a visualization of the following information:

- Algorithms that were selected by Auto AI

- Pipelines that were created applying these algorithms

- Feature transformations that were used by a particular pipeline

The following pipeline leaderboard in Figure 3-5 shows which pipelines produced the best models based on a chosen metric, in this case accuracy.

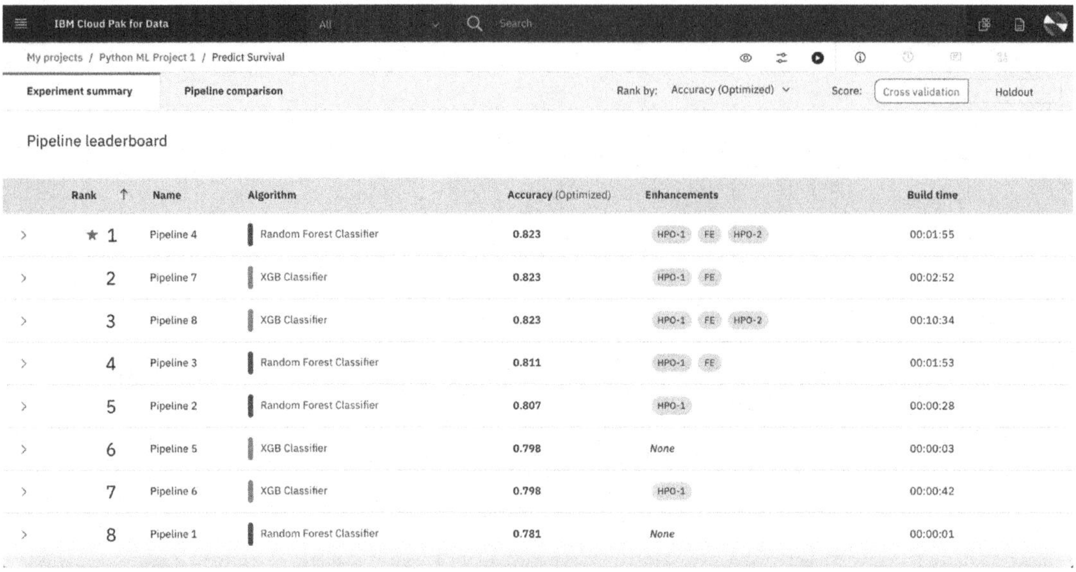

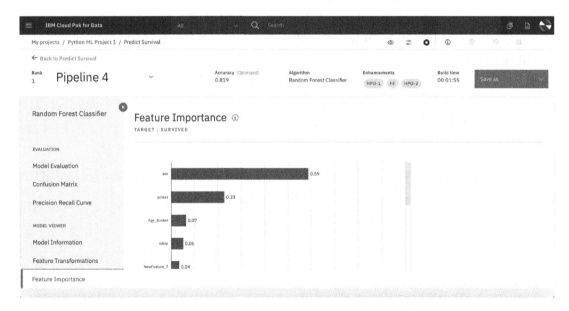

**Figure 3-5.** *Leader Board of Model Training Pipelines Generated by Auto AI*

It is then possible to drill into various metrics of the generated models to explore which pipeline has the best overall result and also understand which features influence the prediction the most, for example, in this case where models were trained with a public data set that has information about Titanic passengers and who survived or did not survive, the most significant feature was a passenger's gender, as you can see in Figure 3-6.

**Figure 3-6.** *Feature Importance Analysis by Auto AI to Help Understand the Model*

# Toward AI Model Eminence

Technical model evaluation metrics such as accuracy, confusion matrix, precision recall, and others are an important factor to consider in choosing models to use.

However, there can be a range of business considerations and applicable laws that ultimately can rule out certain approaches, use of certain features, or may require to intentionally not choose the model with the best technical evaluation metrics.

For example, in models for loan decisions, it would not be acceptable to use the gender of a person or significantly correlated values as a feature influencing whether or not a loan is approved. Depending on explainability or performance requirements, simpler models may need to be selected over models that in principle give better results but would be less explainable or take too long to compute. Depending on the impact of false positives, an overall less accurate model may have to be chosen to minimize the number of false positive predictions, for example, in credit card fraud checks, any false positive would cause enormous customer frustration because they want to pay for something legitimate, but it does not work.

# Deep Learning (DL)

Numerous books, articles, and blogs have been written on DL[8]. Nevertheless, you may still wonder what the difference is between ML, DL, and RL and how these are interrelated to each other. In this book, we don't have the space to extensively cover the DL topic. Rather than describing the theoretical and mathematical underpinning, we intend to enable you to better understand the key characteristics of DL and the differences and the coherence of DL with ML and RL.

## What Is DL?

On the one hand, DL can be seen as a subset of ML, as it delivers true scalable ML by focusing on artificial neural networks (ANNs), which are – as we have seen previously – a particular type of ML algorithms. Implementations of *true scalable* ML solutions depend on special-purpose engines, such as GPUs, FPGAs, or ASICs. Furthermore, DL can be categorized into supervised, unsupervised, and reinforcement learning, similar to the types of ML, which we have discussed at the beginning of this chapter.

---

[8]See [10] for an in-depth treatment of DL and [11] for a more practical guide on DL.

On the other hand, DL can be seen as an evolution of ML, focusing on distinct algorithms that allow for learning in a true sense, meaning with and equally *without* labeled data sets. With ML, we train and retrain an ML model to ensure that accuracy and precision for a pre-defined scenario remain at a certain level, whereas with DL the model can learn "by itself" to become more distinct and accurate over time without manual or tool-based retraining. DL can, for instance, be based on supervised feature learning, leveraging supervised neural networks or supervised dictionary learning, where an initial training may be performed. However, the subsequent increase regarding its relevance and precision of the decision-making process may be done by the ML model itself – based on new data – without human or tool-based intervention. Thus, the DL capabilities are at a different level and are genuinely mimicking the human behavior.

The difference between RL and DL can sometimes be confusing. As you have seen when we discussed the various types of ML at the beginning of this chapter, RL is about learning a policy through rewards and punishments. But RL can be challenging and cumbersome when the data dimension is rather large. This can be effectively addressed by deep reinforcement learning (DRL), where RL algorithms are combined with ANNs or DL algorithms.

Let us now shed some light on ANNs and deep learning networks (DLNs) and clarify for you why we are separating them as different types of neural networks.

# Artificial Neural Networks (ANNs)

ANNs were the first, relatively straightforward neural networks, which loosely model neurons as in a biological brain. Artificial neurons have input connections with weights and output connections and may have a threshold so that they send the output signal only if that threshold is exceeded. ANNs are typically organized in an input layer, hidden layer(s), and an output layer. ANNs are capable of learning various nonlinear functions. In this learning process, weights on connections are adjusted by reducing or minimizing the so-called loss function. The loss function can be viewed as an integral part of using an ANN or DLN algorithm to train theses neural networks.

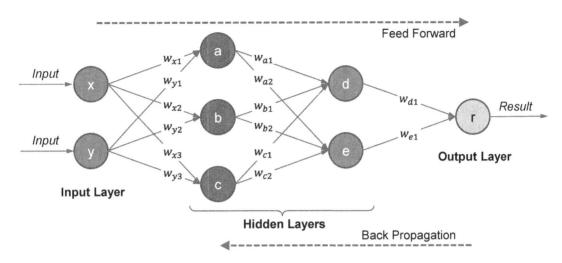

***Figure 3-7.*** *Artificial Neural Network (ANN)*

When we use the term *ANN or DLN algorithm*, we refer to an algorithm that is used to train the neural network. In a way, we can then refer to the trained ANN or DLN as the DL model (to be consistent with our terminology). *Back-propagation* is probably the most popular ANN algorithm, which is actually comprised of a set of training algorithms used for back-propagation. Other ANN algorithms are *radial basis function network* (RBFN), *Hopfield network, multilayer perceptron* (MLP), *stochastic gradient descent,* and so on.

Figure 3-7 is a depiction of a simple feed forward ANN with the input, two hidden, and the output layers. The hidden layers are typically composed of several nodes. In our example, the first layer is composed of nodes a, b, and c, whereas the second layer is composed of nodes d and e. Feed forward refers to the information flowing in only one direction from the input to the output layer. Various back-propagation algorithms (e.g., stochastic gradient decent) can be used to train the ANN, meaning to adjust the various weights $w_{ij}$, which you can see in Figure 3-7.

Applications of ANNs include computer vision, speech recognition, and machine translation. *Deep learning* employs ANNs with multiple hidden layers.

# Deep Learning Networks (DLNs)

ANNs can be seen as the original, classical type of neural networks, which were straightforward and relatively simple. The corresponding ANN algorithms were less sophisticated. As we have mentioned, training the ANN was essentially based on applying various back-propagation algorithms. For today's demanding AI applications, more complex ANNs with a potentially large number of hidden layers are required. However, the resulting number of weights $w_{ij}$ that need to be adjusted for those complex ANNs is becoming incalculably large. An additional problem is the so-called *vanishing and exploding gradient*, which makes the ANN unstable and less capable of learning. This is related to the loss function, which calculates gradients that are too large and therefore not suitable to adequately adjust the weights $w_{ij}$. This phenomenon can even become so prevailing that the ANN is not trainable anymore and therefore not usable.

There are several techniques to address this problem, such as weight regularization or gradient clipping. Nevertheless, more state-of-the-art neural networks have emerged that offer novel and much more innovative approaches to cope with the preceding issues, which we call deep learning networks (DLNs). DLNs and the corresponding DL algorithms are concerned with developing and efficiently training much larger and more complex neural networks, which can process larger volume of data and are suitable for today's challenging AI applications.

Some of the most popular DLNs[9] are *recurrent neural networks* (RNNs), *convolution neural networks* (CNNs), *deep belief networks* (DBNs), deep Boltzmann machines (DBMs), and more. RNNs have a recurrent connection from an individual node to itself, which links the outcome to the input data. CNNs[10] are mainly used for image processing, but also for classification and segmentation. They leverage kernels and filters to train the neural network. A kernel is like an n-dimensional matrix of weights, which is repetitively applied to the input data. A filter is a concatenation of several kernels, where each kernel is assigned to a subset (channel) of the input data.

The emergence and further development and research of these DLNs and the fact that they represent a unique and distinct field in AI lead us to behold DL not to be entirely a subset of ML.

# Decision Optimization (DO)

Decision optimization[11] is a subset of data science techniques used for prescriptive analytics. In contrast to predictive analytics that aims to predict what will happen, prescriptive analytics helps decide what actions to take.

---

[9]See [12] for a high-level overview of some of the most popular DLNs.

[10]See [13] for an in-depth treatment of CNNs.

[11]See [14] for a short overview on decision optimization.

DO analyzes the possible choices and provides the best option from a huge set of alternatives. For example, you do not know if a customer is about to unsubscribe (customer churn) or when an industrial asset might fail, but you can decide whether or not to offer a particular promotion to your customers or decide to run maintenance plans for your assets. These decisions are under your control. They are, however, limited by business constraints, for example, by the maximum promotional budget or by the maintenance crew size. Among all the possible decisions, some are preferred based on objectives you want to optimize. DO prescribes which decision for you to take in such cases.

In Chapter 5, "*From Data to Predictions to Optimal Actions*," we provide an in-depth exploration on the integration of ML with DO[12] and the holistic benefit that can be achieved from combining these two domains.

DO applies to virtually all industries. Example applications include well-known use cases, such as the optimization of supply chains and production planning and scheduling, but also portfolio optimization, predictive maintenance, electricity unit commitment, or even more surprising applications such as price optimization or shelf space optimization.

# Key Takeaways

We conclude this chapter with a few key takeaways, which are summarized in Table 3-1.

---

[12]See [15] for a short overview on the integration of ML with DO.

*Table 3-1.* *Key Takeaways*

| # | Key Takeaway | High-Level Description |
|---|---|---|
| 1 | Machine learning allows training models without explicit programming | Machine learning algorithms build a mathematical model based on sample data, known as "training data," in order to make predictions or decisions without being explicitly programmed to do so |
| 2 | Supervised learning aims to predict | Training data needs to include a desired output that is identified as the prediction target; models are trained to predict output for new input data |
| 3 | Unsupervised learning finds structure in data | Data is analyzed for clusters of similar data |
| 4 | Artificial neural networks can be used for recognition | Modeled after biological neural networks, ANNs can be used for image recognition, speech recognition, language translation, and so on |
| 5 | Decision trees can be used to predict classes or values | Decision trees are constructed based on input variables in order to predict a class (classification decision tree) or continuous value (regression decision tree) |
| 6 | Bayesian networks can predict causes based on observed data | For example, bayesian network models can be built in order to predict the disease that causes observed symptoms |
| 7 | DO is used for prescriptive analytics | In contrast to predictive analytics that focuses on what will happen, decision optimization (DO) focuses on determining decisions/actions |
| 8 | DO can solve complex problems in all industries | Optimization problems are described through constraints, optimization objectives, and input data. DO solves optimization problems, allowing to make optimal decisions to take the best actions |
| 9 | ML plus DO provides enormous value | By combining ML + DO, it is possible to get from data to predictions to optimal decisions and resulting actions. This can drive huge savings relative to brute-force actions based on predictions |

# References

[1]  Wikipedia. *Machine learning,* `https://en.wikipedia.org/wiki/` `Machine_learning` *(accessed April 25, 2020).*

[2]  Shalev-Shwartz, S., Ben-David, S. *Understanding Machine Learning: From Theory to Algorithms.* ISBN-13: 978-1107057135, Cambridge University Press, 2014.

[3]  Alpaydin, E. *Introduction to Machine Learning (Adaptive Computation and Machine Learning series).* ISBN-13: 978-0262043793, The MIT Press, 2020

[4]  Sutton, R. S., Barto, A. G. *Reinforcement Learning: An Introduction (Adaptive Computation and Machine Learning series).* ISBN-13: 978-0262039246, A Bradford Book, 2018.

[5]  Moni, R. Medium. *Reinforcement Learning algorithms — an intuitive overview,* `https://medium.com/@SmartLabAI/` `reinforcement-learning-algorithms-an-intuitive-overview-` `904e2dff5bbc` (accessed April 30, 2020).

[6]  Priyadarshiny, U. DZone. *Introduction to Classification Algorithms,* `https://dzone.com/articles/introduction-to-` `classification-algorithms` (accessed April 29, 2020).

[7]  Hodgson, E. DotActiv. *Clustering Algorithms: Which One Is Right For Your Business?* `www.dotactiv.com/blog/clustering-` `algorithms` (accessed April 29, 2020).

[8]  Silipo, R., Widmann, M. *3 New Techniques for Data-Dimensionality Reduction in Machine Learning,* `https://` `thenewstack.io/3-new-techniques-for-data-dimensionality-` `reduction-in-machine-learning/` (accessed April 30, 2020).

[9]  Vidal, R., Ma, Y., Sastry, S. *Generalized Principal Component Analysis (Interdisciplinary Applied Mathematics).* ISBN-13: 978-0387878102, Springer, 2016.

[10]    Skansi, S. *Introduction to Deep Learning: From Logical Calculus to Artificial Intelligence (Undergraduate Topics in Computer Science)*. ISBN-13: 978-3319730035, Springer, 2018.

[11]    Gad, A. F., *Practical Computer Vision Applications Using Deep Learning with CNNs: With Detailed Examples in Python Using TensorFlow and Kivy*. ISBN-13: 978-1484241660, Apress, 2018.

[12]    Gupta, R. Medium. *Towards data science. 6 Deep Learning models — When should you use them?* `https://towardsdatascience.com/6-deep-learning-models-10d20afec175` (accessed May 1, 2020).

[13]    Khan, S., Rahmani, H., Shah, S. A. A., Bennamoun, M. A *Guide to Convolutional Neural Networks for Computer Vision (Synthesis Lectures on Computer Vision)*. ISBN-13: 978-1681730219, Morgan & Claypool Publishers, 2018.

[14]    IBM. *Decision optimization*, `www.ibm.com/analytics/decision-optimization` (accessed April 29, 2020).

[15]    Chabrier, A. Medium. *Combine Machine Learning and Decision Optimization in Cloud Pak for Data*, `https://medium.com/@AlainChabrier/combine-machine-learning-and-decision-optimization-in-cloud-pak-for-data-60e47de18853` (accessed May 1, 2020).

# PART II

# AI Deployment

# CHAPTER 4

# AI Information Architecture

In this chapter, you will learn about the specific role of information architecture (IA) to deliver a trusted and enterprise-level AI foundation. As an introduction into this topic, we briefly review key aspects of an information architecture (IA) and highlight the logical and physical IA components in the context of AI. These are important to the reader in order to fully understand the impact of AI on an existing information architecture. Any architecture needs to be underpinned with products and offerings. We learn about the key components of AI information architecture and their role in enterprise suitability. We conclude with use cases that illustrate AI in information architecture.

At IBM, we have helped some of our customers to begin their AI journey simply by downloading and using some freely available open source packages and libraries, for instance, Apache Spark libraries. There needs to be enterprise-wide guidelines available; otherwise, this may result in different open source libraries used by different departments or organizations within the enterprise. As those packages and libraries are not integrated and interoperable, AI artifacts, that is, ML or DL models, have different formats and cannot be easily exchanged and deployed across organizations and different platforms. Any AI endeavor requires more than downloading a few freely available open source ML or DL libraries and runtime engines. You should take a holistic, enterprise-wide approach and plan for collaboration, integration, specialized engines, open source, enterprise-scale offerings, and so on.

Furthermore, each department or organization is struggling to access and transform the required source data for adequate consumption in a particular application or use case context. Governance of access and usage of AI artifacts, which we will talk about in Chapter 8, "*AI and Governance*," is again left to individual departments and organizations. Thus, the lack of an enterprise-wide information architecture does often result in inconsistent and clumsy shadow IT implementations across the enterprise.

© Eberhard Hechler, Martin Oberhofer, Thomas Schaeck 2020
E. Hechler et al., *Deploying AI in the Enterprise*, https://doi.org/10.1007/978-1-4842-6206-1_4

Other organizations may start their AI journey by deploying a commercially available AI toolset, that is, an ML or DL offering, such as IBM Watson Machine Learning (ML) for z/OS that is comprised of tens or even hundreds of libraries and algorithms, some runtime engines, and components to store and manage AI artifacts. However, source data discovery and provisioning into the AI toolset with the required data quality remains a challenging endeavor. Getting your people trained and accustomed to such a tool stack is yet another challenge. Access and transformation of source data into the required data and feature format is often incomplete or missing at all. Subsequent data synchronization and organization tasks and data privacy and protection needs are often outside the scope of AI tools. Furthermore, the integration of these AI tools into an existing IT and business landscape is accompanied with unforeseen organizational and technical hurdles, often turning AI projects into a frustrating experience. Thus, the lack of an enterprise-wide information architecture does result in a fragmented AI infrastructure that makes enterprise-scale AI projects and use cases challenging and risky undertakings.

There is *no AI without IA*[1] may sound like yet another buzz phrase. However, this chapter provides further justification to adequately invest in your IA as you move up your AI ladder.

# Information Architecture – A Short Review

Since information-intense business challenges exist already for some decades, there is a rather large body of knowledge and publications available that are related to information architecture[2]. This section provides a very short review of the terms information architecture and enterprise information architecture (EIA) and defines the most important aspects, such as enterprise architecture layers, the concept of a reference architecture (RA), and EIA reference architecture work products. We furthermore provide a short review of key architecture models and methods, such as the information maturity model. This short review is of generic nature and should enable the reader to understand the terminology and concepts, which we elaborate on in the subsequent sections of this chapter.[3]

---

[1]More motivation for an information architecture within an AI context can be found in [1].

[2]See [2] and [3] for more information on EAI and related concepts.

[3]We'd like to refer the reader to [2] and [3] and encourage the further study of IA concepts to get a broader appreciation and understanding of IA and its relevance for an AI strategy.

# Terminology and Definitions

We begin our review with a definition of an enterprise architecture, as this provides a means to position the terms IA and EIA. The Open Group Architecture Framework (TOGAF),[4] a standard of The Open Group, is one of the most comprehensive enterprise architecture methodologies and frameworks[5]. According to the TOGAF standard, the objective of an enterprise architecture *"is to optimize across the enterprise the often fragmented legacy of processes (both manual and automated) into an integrated environment that is responsive to change and supportive of the delivery of the business strategy."* An enterprise architecture provides a framework to facilitate the alignment between the business strategy, IT strategy, and the IT implementation. As such, it embraces various domains that address the enterprise strategy, the needs from the business and application domains, the information (or data) domain, and the infrastructure.

The following are the architectures that are part of the TOGAF:

- Business architecture

- Data architecture

- Application architecture

- Technology architecture

According to TOGAF, the data architecture *"describes the structure of an organization's logical and physical data assets and data management resources."* For our purpose and scope of this book to demystify AI, the information architecture provides the foundational information-relevant concepts, methodologies, and frameworks to guarantee responsiveness and trusted information insight that the business requires from its information layer. Broadening and expanding the data architecture, the information architecture leverages the information-centric systems and components of the IT environment and defines its relationship to the business objectives.

The information architecture furthermore describes the principles and guidelines that enable consistent implementation of information technology solutions, how data and information are both governed and shared across the enterprise, and what needs to be done to gain business-relevant trusted information insight. We will come back to this particular aspect in Chapter 8, *"AI and Governance."*

---

[4]See [4] for more information on The Open Group Architecture Framework (TOGAF) Version 9.1.
[5]The interested reader may also look for other standards and frameworks.

The following are some examples of the core principles that guide an information architecture:

- **Access and the flow of information**: Information services should provide unconstrained access to the right users at the right time and provide means to facilitate the required information flow.

- **Service reuse**: Facilitates discovery, selection, and reuse of information-related services and – whenever possible – encourages the use of uniform interfaces.

- **Information governance**: An adequate information architecture and the corresponding information technology should support the efficient execution of an information governance strategy.

- **Standards**: A set of coherent standards for data and technology should be defined to promote simplification across the information infrastructure.

The enterprise information architecture (EIA) is the framework that defines the information-centric principles, architecture models, standards, methods, and processes that underpins the enterprise architecture. The EIA addresses the information technology decisions across the enterprise and relevant business organizations. Thus, the EIA translates the business requirements into informational strategies and addresses the entire information supply chain from available and required data components to the derived insight that is consumed by business applications. The use of "*enterprise*" in the EIA term adds the enterprise-wide business context to the definition of information architecture described in this section.

Before we elaborate on IA methods and models, let us introduce the term reference architecture[6]. A reference architecture (RA) is an enterprise architecture that addresses the specific needs of a particular domain, industry, or theme, such as a logical data warehouse (DWH), or the automotive industry, or AI. It is based on the principles of an enterprise architecture framework, but is adjusted to accommodate the specific requirements of that domain, industry, or theme. Thus, we speak of an AI reference architecture (AIRA), or an enterprise information architecture reference architecture (EIARA). Ideally, any RA should be based on a number of concrete client engagements

---

[6]See [5] for a short definition of the term reference architecture.

and solution deployments. It should be underpinned with best practices that are harvested from those engagements and deployments. In the following sections, we describe key aspects of an IA for AI, which we may call AI information reference architecture (AIIRA) or more simply AI information architecture (AIIA).

# Methods and Models

In past decades, many enterprise architecture methodologies have been developed. In this short introduction, we continue to focus on the TOGAF. The Architecture Development Method (ADM)[7] of the TOGAF is comprised of various methods to develop architectures. One method is the Information Systems Architectures – Data Architecture ADM *"that enables the Business Architecture and the architecture vision, while addressing the request for architecture work and stakeholder concerns."* It describes in detail the enterprise architecture development cycle, including its iterative approach, the required adaptation, and decisions to be addressed that relate to the specific scope and architectural assets of the enterprise architecture.

There are many enterprise and information architecture guidelines and principles, architecture styles and patterns, best practices, and models available, which are either of a more generic nature or contained within a reference architecture that relates to a specific industry (insurance, automotive, etc.) or domain (logical DWH, AI, etc.). In this short review, we briefly elaborate on the IA information maturity model. In subsequent sections of this chapter, we relate this information maturity model to the specific AI domain.

An information maturity model is a model or technique to assess the maturity of the IA and the maturity of transforming data and information into business-relevant insight. It can be seen as a technique or a set of imperatives to develop an IA that addresses the following areas:

1. **Reducing the time needed to access information**: Accessing and provisioning data and information in a timely manner to consuming business systems and applications.

2. **Reducing information complexity**: Addressing the diversity, complexity, and variety of formats of data from structured (e.g., relational) to unstructured (e.g., text, social media, videos).

---

[7]See [6] for more detailed information on the Architecture Development Method (ADM) of the TOGAF.

3. **Lowering costs through an optimized infrastructure**: Reducing cost through a simplified IA that is agile and adjustable to changing business needs and frequently occurring disruptions.

4. **Gaining insight through analysis and discovery**: Being able to discover, analyze, and accelerate inference and insight is a key need for all line-of-business organizations.

5. **Leveraging information for business transformation**: Continuously infusing data and information to facilitate business transformation and to enable the implementation of emerging business use case scenarios.

6. **Gaining control over master data**: Managing master data (e.g., customer, business partner, employees, products, and services) and reference data through adequate enterprise MDM systems[8].

7. **Managing risk and compliance via a single version of the truth**: Addressing compliance to an ever-increasing number of business regulations and furthermore understanding and mitigating risks.

As mentioned before, our pending task is to adjust and map the preceding areas (and similar areas) to the specific needs of AI to develop an AI information architecture.

# Enterprise Suitability of AI

In the previous sections, we have briefly reviewed key concepts of an enterprise and information architecture. This section describes the suitability and implications of an IA in the context of AI. We elaborate on the influencing factors of AI that drive the need for a comprehensive AI information architecture.

# Relevance of Information Architecture for AI

As organizations are developing their AI strategy and are increasingly using AI, machine learning, and deep learning, the need for adjusting and improving their existing IA becomes obvious. As you have seen in Chapter 1, "*AI Introduction*," new use cases

---

[8]See [7] for a comprehensive treatment of enterprise MDM systems.

and usage patterns of data and information are emerging, and new artifacts, such as analytical models, will be developed.

In addition, innovative and efficient collaborative models of various roles and responsibilities across various organizations have to be implemented. Business users, data scientists, data engineers, and IT operations specialists have to effectively collaborate in order to exchange and govern new types of artifacts; to provision and transform data into new consumption patterns; to introduce new components for ML and DL model development, learning, deployment, and management; and to deploy and operationalize, for instance, ML and DL models in conjunction with new types of applications, such as real-time transactional scoring or real-time customer classification.

Table 4-1 outlines the relevance of an information architecture for AI and lists the imperatives and needs that drive an AI information architecture. This list may neither be complete, nor has it been arranged in a prioritized order. Obviously, there is some overlap and correlation among some of the imperatives and needs.

***Table 4-1.*** *Imperatives and Needs for an AI Information Architecture*

| # | Imperatives | High-Level Description |
|---|---|---|
| 1 | New and emerging use cases | As has been described in Chapter 1, "*AI Introduction*," numerous use cases and emerging business areas, such as driver assistance systems and autonomous driving, managing a paradigm shift from existing rule-based fraud discovery systems to using DL and ML, integrating natural language processing (NLP) to improve the human-computer interaction, and so on, are driving the AI information architecture |
| 2 | New usage patterns of data and information | For AI-related scenarios, data and information needs to be processed, provisioned, and governed in new ways. Processing may require to label data or to discover correlation among hundreds of data components. Provisioning may require transactional data and other predictors to be transformed and integrated for real-time or low-latency scoring. Governance[9] may have to be adjusted to, for instance, address new ML- and DL-related artifacts |

(*continued*)

---

[9]See Chapter 8, "*AI and Governance*," for more detailed information.

***Table 4-1.*** (*continued*)

| # | Imperatives | High-Level Description |
|---|-------------|----------------------|
| 3 | Accommodating new artifacts and standards | AI requires the development, integration, and management of new artifacts, such as ML and DL models, Jupyter Notebooks, metadata, and so on. New standards such as the Open Neural Network Exchange (ONNX) format[10] and the existing Predictive Model Markup Language (PMML) standard[11] need to be taken into consideration |
| 4 | Innovative and efficient collaboration needs | AI creates the demand for collaboration across various user roles and responsibilities, such as business analysts, business users and application developers, data scientists, data engineers, and DevOps and IT operations specialists. This relates to innovative techniques to facilitate the exchange and management of requirement statements, resources, new artifacts, and DevOps scenarios |
| 5 | New consumption patterns of insight | AI methods generate new ways to gain insight, which demand new insight consumption patterns and inference techniques and capabilities to be supported by the AI information architecture. For instance, ML and DL models are not static: ML models can be retrained to maintain a certain level of accuracy and precision, and DL models can be learned to continuously improve and validate the relevance of insight. The AI information architecture needs to support monitoring, retraining, and continuous learning of ML and DL models |

(*continued*)

---

[10]See [8] for more details on the ONNX format.

[11]See [9] for more details on the PMML standard.

***Table 4-1.*** (*continued*)

| # | Imperatives | High-Level Description |
|---|---|---|
| 6 | Distributed development and deployment patterns | The preceding user roles and responsibilities may be performed on different platforms, such as open source Hadoop-based data lakes (e.g., Hortonworks HDP), distributed or IBM Z (mainframe systems), and private, public, or hybrid cloud environments. The source data may originate and reside on different platforms. Subsequently, there is a need to allow for development, training, test, and validation, for instance, of ML and DL models on one platform and to facilitate the deployment and operationalization including scoring on a different platform. The AI information architecture needs to allow the seamless distribution of all tasks across these platforms |
| 7 | Deployment and operationalization aspects | Today, the AI operationalization[12] challenges represent a huge pain point for a number of organizations. For instance, the need to provide real-time decisions and scoring requires data provisioning and preparation to be done in a timely manner, which may include to access, assemble, and prepare data residing in a variety of source systems |
| 8 | Integration of specialized engines and accelerators | Specialized engines, such as a graphics processing unit (GPU), application-specific integrated circuit (ASIC) engines, or systems that are optimized for AI usage, such as the IBM PowerAI system[13], have to be integrated into the IT landscape, requiring adjustments of the existing IA |
| 9 | Addressing different data access and transformation needs | Depending on the use case, new data from new sources or devices may have to be captured and processed in real time (e.g., sensor data, voice data, etc.). Even existing source data, such as records from financial transactions, have to be accessed in real time and prepared for scoring |

(*continued*)

---

[12]See Chapter 6, *"The Operationalization of AI,"* for more details of this topic.

[13]See [10] for more information on the IBM PowerAI system.

***Table 4-1.*** (*continued*)

| # | Imperatives | High-Level Description |
|---|---|---|
| 10 | Easy data search and data exploration capabilities | Especially data exploration to identify new patterns in data or to discover correlation between different KPIs or source data points (e.g., call center voice data and the transactional data for a particular customer) is an essential task, for instance, to identify relevant features or predictors for ML model development. The AI information architecture needs to facilitate simple data search, access, and exploration capabilities |
| 11 | Introducing learning and reasoning into the existing IT and business landscape | New aspects, such as learning, complementing understanding with reasoning, or training with no initial labeled data sets (meaning autonomous learning[14]), represent new challenges and opportunities. Additional scenarios, such as learning with zero input (learning from scratch) or autonomous driving in weird situations, represent future challenges not only for the AI information architecture but the overall enterprise architecture |
| 12 | Integrating voice and other interfaces | Voice interfaces improve the human-computer interaction through NLP. Voice interfaces, eye retina scanning, and facial expression recognition to detect sentiments or to calculate a mood score of a customer or to discover the sleepiness of a car driver will thus allow the system to gain a much broader and deeper understanding about a customer or a driver. These interfaces and scanning systems and their corresponding data need to be taken into consideration when developing an AI information architecture |

As mentioned earlier, this list could very well be extended. For instance, commonly and predominantly used languages, such as Python, Scala, and R, need to be taken into consideration when developing an AI information architecture, and open source components, runtime engines, and libraries, such as Apache Spark, scikit-learn, TensorFlow, and Caffe – just to name a few – need to be integrated. Enterprise suitability, security, scalability, cloud readiness, and availability are additional key imperatives.

---

[14]See Chapter 13, *"Limitations of AI,"* for more information on autonomous learning.

For some of the aspects listed in the preceding table, we will elaborate more on in subsequent sections of this chapter.

## Information Architecture in the Context of AI

Figure 4-1 depicts the coherence between enterprise architecture, information architecture, and the AI information architecture.

The AI information architecture can be seen as an IA, where AI-specific needs and imperatives as listed in Table 4-1 and reference architecture principles as they relate to the AI domain are applied. Since we now understand the driving aspects and motivation to describe an AI information architecture, we intend to transform the preceding imperatives into a list of technical capabilities that are needed from a conceptional perspective.

Figure 4-2 is a depiction where these capabilities are shown conceptually through an architecture overview diagram (AOD). This AOD can be leveraged in order to identify candidates for AI information architecture building blocks[15].

Depending on the AI strategy and chosen use cases, not all of these building blocks may have to be implemented right from the beginning. However, the AI information architecture should be designed with scalability, extensions, and enhancements in mind. It should be agile in order to address future requirements.[16] AI tools and offerings should underpin the need for agility of the AI information architecture.

---

[15]TOGAF 9.2 defines an architecture building block (ABB) and characterizes ABBs through a well-defined specification with clear boundaries.

[16]See [24] for more information on building trust in a smart society.

**Addressing the enterprise strategy, including business and IT**

- Providing a framework to facilitate the alignment between the business strategy, IT strategy, and the IT implementation
- Embracing business, application domains, information, and infrastructure

*Enterprise Architecture*

**Focus on physical and logical data and information assets**

- Information-relevant concepts, methodologies and frameworks
- Information-centric systems and components
- Principles and guidelines that enable consistent implementation of information technology solutions
- Information Maturity Model

*Information Architecture*

**Specific needs of AI to develop an AI Information Architecture, e.g.**

- Introducing machine learning (ML) and deep learning (DL) methods
- Cataloging and governing ML and DL artifacts
- Deploying and operationalizing ML and DL models
- Exchanging artifacts across IT platforms and business systems
- Ensuring model accuracy and precision ot their entire lifecycle
- Infusing insights from ML models with business via reporting tools

*AI Information Architecture*

***Figure 4-1.*** *The AI Information Architecture in Context*

Certain aspects and capabilities that make up a pervasive AI information architecture are not taken into consideration, such as specific data lake aspects and integration aspects with traditional enterprise BI and DWH systems. The role of MDM systems to manage core information will be briefly elaborated on in the following section.

We have categorized the core architecture building blocks (ABBs) of the AI information architecture into the following six layers or categories, whereby the ABBs of the information governance and information catalog[17] provide services to all other five layers:

1. Data sources

2. Source data access layer

3. Data preparation and quality layer

4. Analytics and AI layer

5. Deployment and operationalization

---

[17]See Chapter 8, "*AI and Governance,*" for more details.

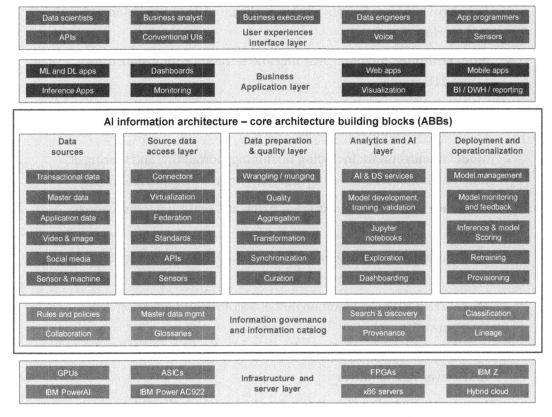

*Figure 4-2.* *AI Information Architecture Overview Diagram (AOD)*

The following is a subset of some of these building blocks[18] as they have been depicted in the preceding AOD:

- **Source data access**: Including data connectors, virtualization, and federation

- **Data preparation**: Including wrangling/munging, aggregation, and transformation

- **Data quality**: For instance, to ensure trusted input for downstream data consumption

- **Data exploration**: Exploratory data analysis including correlation discovery, data mining, and so on

---

[18]We make no claims for this list to be complete; adjustments and extensions may be required depending on the scope and use case of a particular project.

- **Data provisioning**: Specifically for applications to enable inference and scoring

- **Model training**: Including learning and retraining, for instance, for ML and DL models

- **Model validation and testing**: To ensure highest accuracy and precision of ML and DL models

- **Model management**: Including versioning, portability, and scoring of different model versions using different runtime engines, and so on

- **Inference and model scoring**: For batch and real-time transactional scoring

- **Model monitoring and feedback**: Regarding model accuracy and precision over their entire life cycle

- **Model retraining**: To maintain model accuracy and precision over their entire life cycle

- **Specialized engines and accelerators**: For example, GPUs, ASICs, IBM PowerAI, and others

- **Information catalog**: For all AI-related assets, including ML and DL models

- **Information governance**: Including policies, rules, lineage, provenance, and others

The AOD as depicted in Figure 4-2 is the highest-level abstraction of the AI information architecture. It is based on the scope of the conceptual level of a reference architecture, which is closest – in our case – to the AI business definitions and processes. A reference architecture consist of the following three levels:

1. **Conceptual level**: Described as the conceptual architecture

2. **Logical level**: Described as the logical architecture and component model

3. **Physical level**: Described as the operational model

In addition to the AOD, this conceptual level, which is also called conceptual architecture, provides a high-level architecture overview and lists the required capabilities and ABB candidates.

As described previously, there are two remaining levels of the reference architecture, the logical level, with its logical architecture and conceptual model, and the physical level, with its operational model. Some aspects of the logical level of the reference architecture will be used in the next section, where we will describe the AI information architecture in the context of the ML workflow, especially as it relates to the component relationship diagram and the ML-related data flow.

Some aspects of the physical level of the reference architecture will be used in Chapter 6, "*The Operationalization of AI*," where we elaborate on AI operationalization and deployment aspects.

In order to stay focused on the core ABBs of the AI information architecture, we neither elaborate any further on the user experiences and interface layer nor on the business application layer. The various personas, such as business executives and decision makers, data scientists, business analysts, data engineers, and application programmers, will be referenced again in the next section on the ML workflow.

Although very important for a holistic AI solution, the infrastructure and server layer with its general-purpose systems and its special-purpose engines, such as GPUs, ASICs, FPGAs, IBM PowerAI, and IBM Power System AC922[19], are also out of scope of this book.

# AI Information Architecture and the ML Workflow

The previous section has introduced quite a number of AI information architecture capabilities and corresponding ABBs. In this section, we describe the relationship and interaction of some of these building blocks or components specifically in the context of the ML workflow.

The following are some of the key capabilities that specifically relate to the ML workflow:

- **Model portability**: Develop anywhere and deploy anywhere to enable transparency across platforms

- **Model monitoring**: Feedback from inference to understand degradation

- **Model accuracy**: Retrain models to sustain change and to guarantee accuracy

---

[19]See [10] for more information on the IBM Power System AC922.

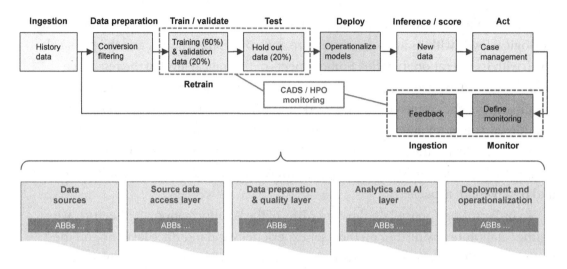

***Figure 4-3.*** *AI Information Architecture and the ML Workflow*

Figure 4-3 depicts the ML workflow in the context of the core ABBs of the AI information architecture. In this section, we do not provide a detailed explanation of the ML workflow itself; the goal rather is to relate the ML workflow to the AI information architecture.

The ML workflow starts with the access and ingestion of source and historical data. Depending on the use case and corresponding data sources, this will require various capabilities, such as data connectors (e.g., using JDBC), data virtualization, and others. Sensor or voice data may be required as well for subsequent steps, such as for developing and training of ML or DL models.

The data preparation and quality improvement phase is typically the most intensive and lengthy step in the entire ML workflow; it can very well consume more than 80% of the effort and overall elapsed time. Data wrangling or munging[20], in conjunction with data transformation, aggregation, quality improvements, filtering and curation, and so on, will ensure that source data, including sensor, voice, text, or image data, is trusted, complete, consumable, and "ready" for AI.

ML and DL model training, validation, and testing[21] are well-known steps as part of the model development process. In this book, we won't describe any details of these steps.

---

[20]Data wrangling or munging is the transformation of raw and often incomplete and nonconsumable data into a consumable format for any downstream data processing step.

[21]See [11] for more information on ML model training, validation, and testing.

For most business scenarios, the deployment and operationalization of models, as well as the inference and scoring of ML and DL models, represents a real challenge. This is mainly because during the ML and DL model development phase, a variety of data sources have been taken as a base to identify suitable predictors and features. The resulting model(s) may address perfectly well the business requirements.

However, after the deployment, the inference and scoring applications depend on the timely provisioning of new data in the "right" format. This new data, however, still comes from the same variety of source data systems and may have to be assembled in real time or with low latency to serve as input for inference and scoring applications. This process in the ML workflow requires a comprehensive and flexible AI information architecture that is comprised of ABBs, such as the provisioning, inference, and model scoring ABBs.

We will elaborate further on the operationalization and deployment challenges in Chapter 6, "*The Operationalization of AI.*" Once analytical insight, such as a score for a fraudulent transaction or customer classification for targeted marketing campaigns, is available, case management applications may be used to determine the corresponding action based on this insight. We may very well relate this particular step in the ML workflow to the business application layer.

ML and DL models may become less accurate and precise over time. This is because fraud patterns, customer behavior, or business imperatives may change as well over time. Thus, the accuracy and precision of ML regression or DL models have to be monitored. Even clustering and classification may have to be redone, if the underlying assumptions or data points will be changed. For ML regression models, retraining has to be facilitated depending on defined thresholds. In some cases, existing ML and DL models have to be adjusted in terms of adding additional predictors or evaluating additional algorithms that may yield better accuracy or precision. The ML workflow as depicted in Figure 4-3 illustrates these tasks. The AI information architecture needs to support the monitoring and feedback especially of ML regression model accuracy and precision, for instance, through a set of RESTful APIs or convenient GUIs. Most of the ABBs of the deployment and operationalization category address these tasks.

Some of the IBM offerings include innovative capabilities to optimize performance and parameter setting during the model development cycle. The cognitive assistant for data scientists (CADS) and hyperparameter optimization (HPO)[22] ensure performance

---

[22]See [12] for more information on cognitive assistant for data scientists (CADS) and hyperparameter optimization (HPO).

optimization for ML model development. CADS makes it easier for a data scientist to identify the correct algorithms and develop the best model. This process is usually done through lengthy trial-and-error testing. HPO helps data scientists identify and select the best parameters. This feature, in turn, helps them fully optimize the predictive capabilities of their models. In the following section where we address the mapping to IBM offerings, we will come back to the CADS and HPO functionality.

# AI Information Architecture for Any Cloud

Cloud computing is around since the 1960s. The concept of using remote servers, middleware software, data, and applications without installing and hosting infrastructure, software, and applications has undergone a major boost with the emergence of the Internet in the 1970s and 1980s. Today's leading enterprises are already using cloud services, such as infrastructure as a service (IaaS), platform as a service (PaaS), and software as a service (SaaS) in public, private, or hybrid cloud deployment models.

However, AI as a service (AIaaS) seems to be a lot less mature in terms of offerings, platforms, and tools available from vendors. AI as a service requires various capabilities that extend from data- and information-related services to specific AI services on any cloud. As a consequence, a state-of-the-art information architecture is required that enables and supports AI as a service.

The AI information architecture as described earlier should be implemented as an integrated and modular set of capabilities via infrastructure, tools, offerings, and AI services that underpin a cloud platform for data and AI. This cloud platform for data and AI includes an infrastructure, server layer, and the core ABBs of the AI information architecture, as depicted in Figure 4-2, which is either provided as an integral part for a private cloud implementation or provided by a public cloud service provider.

The IBM Cloud Pak for Data System[23] is such an offering that addresses a vast set of needs to enable data and AI on any cloud environment. A high-level description of the IBM Cloud Pak for Data is provided further down in this chapter.

---

[23]See [13] for more information on the IBM Cloud Pak for Data System.

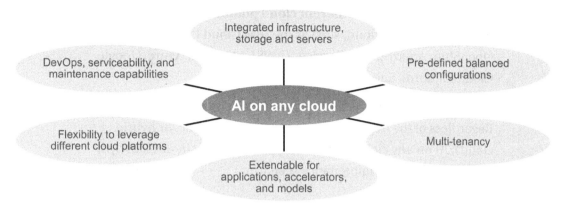

***Figure 4-4.*** *Imperatives for AI on Any Cloud*

In addition to the already identified AI information architecture capabilities, AI on any cloud needs to address the additional imperatives as depicted in Figure 4-4, which are essential to deliver a comfortable experience and fast deployment of AI services in conjunction with information-related services:

1. **Integrated infrastructure, network, storage, and servers**: Especially for a private cloud deployment, all required infrastructure components, network elements, storage, general-purpose servers, and specialized engines, such as GPUs, FPGAs, ASICs, the IBM PowerAI for DL, or the IBM Power System AC922, need to be pre-integrated.

2. **Pre-defined balanced configurations**: Depending on the workload and tasks to be performed, data volume, number of users, number and processing complexity (in terms of GPUs needed) of ML and DL models to be trained and managed, and the specific use case demand, the system needs to be delivered with pre-defined T-shirt-size configurations that users can conveniently choose from.

3. **Multi-tenancy**: To enable multiple applications and users to operate within a shared environment in a secure fashion.

4. **Extendable for applications, accelerators, and models**: The AI information architecture for AI on any cloud should either anticipate the inclusion of industry-specific applications, ML and DL models, and accelerators or should allow the deployment and import of assets from ISVs.

73

5. **Flexibility to leverage different cloud platforms**: AI should be deployable on any cloud, such as the IBM Cloud Pak, Red Hat OpenShift, Amazon Web Services (AWS), Microsoft Azure, Google Cloud Platform, and so on.

6. **DevOps, serviceability, and maintenance capabilities**: Regardless of the deployment model (private, public, or hybrid cloud), AI on any cloud needs to support state-of-the-art DevOps, serviceability, and maintenance scenarios.

# Information Architecture for a Trusted AI Foundation

As we have seen in this chapter, there are a number of core ABBs required to develop an AI information architecture. In this section, we enlarge upon the following aspects of the AI information architecture, which are critical to deliver a trusted AI foundation:

- Data discovery and trustworthiness of data

- Data transformation and synchronization

- Data exploration to gain relevant insight

- Data provisioning for relevant and timely inference

- Role of master data management (MDM) for AI

These capabilities are already well-known and play a vital role in many information reference architectures that are geared toward a particular domain, industry, or theme, such as a logical DWH, or an industry (e.g., automotive), or data lake implementations. However, we refer and limit the discussion specifically to AI.

# Data Discovery and Trustworthiness of Data

Data, information[24], and artifacts are typically stored in different formats and scattered throughout the entire enterprise in different silos, which makes search and discovery a challenging endeavor. For some use cases, data and information may even be external to the enterprise, for instance, weather data, financial and stock trading data, blacklisted data (e.g., credit card numbers), social media data, third-party data, and so on. For all these data sources, the data credibility and trustworthiness, such as completeness, accuracy, quality, currency (actuality), provenance, and so on, is pivotal for all AI scenarios.

The development of ML and DL models, including the training and evaluation/test phases, requires enterprise-ready search and discovery capabilities for data scientists to identify relevant data (internal and external to an enterprise) to address a particular business problem. These search and discovery capabilities need to cope with various data and information formats.

Credibility and trustworthiness needs to be guaranteed not only for the development of models but also for inference and scoring. This represents a challenge, especially if timeliness is required, for instance, for real-time scoring to prevent fraud or to recommend a next best action.

We like to point out that AI – specifically ML methods – could and should be used to increase the credibility and trustworthiness of data. This particular opportunity, however, won't be further elaborated on in this book.

# Data Transformation and Synchronization

Source data is rarely available in a format that is consumable for exploration and ML or DL model building, training, and validation/testing, nor is it adequate for timely inference. Thus, data transformation, aggregation, and synchronization is required to ensure adequate consumability of source data for AI scenarios.

---

[24]We are using the term "information" as derived insight from data in a defined business context. Examples of "information" are business reports, or a fraud score, that are derived from "raw" transactional data.

Although we are using the same terminology as for building a DWH, these tasks have the following implications and aspects that are very specific to AI:

- **Data transformation is intertwined with gaining insight**: Business agility and the need for continuous business adaptation and optimization require a linkage and intertwining of data transformation tasks while ML and DL models are being developed. While additional and deeper insight is gained, transformation will reoccur and needs to be adjusted over time. In the context of AI, this is an iterative process, not a stable or static process comparable to the well-known ETL processes for building, for instance, an enterprise DWH.

- **Learning and monitoring continuously impacts data transformation**: Due to continuously monitoring the accuracy and precision of ML and DL models, and the ongoing learning of DL models, the data pipeline and transformation need to be continuously adjusted. For instance, new features need to be taken into consideration to allow for ML and DL model adjustments and improvements.

- **Data transformation itself is impacted by ML**: Contrary to traditional ETL processes for DWH systems, data transformation needs are infused with ML techniques, for instance, to perform ML-based data exploration or predictor (feature) development. In other words, data transformation is largely integrated with the development of data pipelines.

The preceding characteristics require the corresponding ABBs of the AI information architecture to be uniformly integrable and assimilable, for instance, in the entire ML workflow.

# Data Exploration to Gain Relevant Insight

AI is commonly associated with predictive modeling and classification. However, data exploration and non-supervised learning methods, such as data mining tasks, principal component analysis (PCA)[25], or clustering, constitute a large part in any AI endeavor, possibly consuming even more than 80% of the overall time. The corresponding ABB "exploration" within the analytics and AI layer of our AI information architecture needs to support these AI-specific tasks.

Data exploration is typically provided as an integral set of capabilities within any ML and DL toolset. In the next two sections, where we introduce the IBM Cloud Pak for Data as an exemplary offering and discuss the mapping of the AI information architecture to some sample vendor offerings, we will revisit data exploration as part of those offerings.

# Data Provisioning for Relevant and Timely Inference

Once ML or DL models have been developed and deployed, new data needs to be provisioned for relevant and timely inference. This is usually the collaborative responsibility of IT operations specialists and application programmers. Application programmers develop, for instance, scoring applications that require predictors in a particular format as input, whereas IT operations specialists need to provision these required predictors based on raw transactional data and possibly other data sources that may even be external to the company.

This is a collaborative task, where data provisioning relates significantly to the previous data preparation and quality step, including corresponding data transformation, quality, and aggregation tasks.

Contrary to traditional ETL, where this is commonly done as batch processes overnight, the data provisioning for AI requires timely inference. This is why the AI information architecture needs to rely on the data provisioning tasks to be underpinned with data streaming and acceleration and is therefore associated with an infrastructure and server layer, which is comprised of special-purpose engines, such as GPUs, ASICs, FPGAs, and so on.

---

[25]Principal component analysis (PCA) is a method to reduce a large set of possibly correlated components or dimensions (predictors, features) to a (possibly even smaller) set of uncorrelated dimensions called principal components. Please see Chapter 3, "*Key ML, DL, and DO Concepts*," for more details.

# Role of Master Data Management (MDM) for AI

Managing core information, such as customer, employee, business partner, product or services data, and others, and integrating trusted master data in any business process can be a challenging task, which is usually addressed by integrating MDM[26] customer data integration (CDI) and product information management (PIM) systems into the business and IT landscape.

How AI and MDM influence and enrich each other is an interesting topic, which we elaborate further on in Chapter 9, "*Applying AI to Data Governance and MDM*," where we concentrate on *ML-based matching for MDM*. In the context of the AI information architecture, we list just a couple of specific imperatives related to MDM:

- **MDM for the ML workflow**: Trusted core information needs to be incorporated into the ML workflow, which includes ML and DL model training, validation/testing, as well as inference. For instance, customer data needs to be available in real time, for instance, for inference and scoring and vice versa: CDI systems also need to be updated with new customer insight, for instance, related to product and services preferences of individual customers.

- **AI to augment MDM scenarios**: As marketing campaigns and next best actions, and as products, services, and offers that are recommended to customers, will become more targeted for individual customers, AI with ML and DL methods needs to become an integral part of MDM systems in order to augment and improve MDM-centric scenarios. AI can also improve MDM, for instance, by predicting more relevant core information components or improve accuracy of matching core information segments.

# Mapping to Sample Vendor Offerings

In order to fully take advantage of AI, most leading vendors provide comprehensive information architecture platforms with corresponding services. In this section, we feature some examples from IBM, Amazon, Microsoft, and Google.

---

[26]See [7] for more information on enterprise master data management.

# IBM Cloud Pak for Data

In this section, we describe – as a possible example – how the AI information architecture relates to the IBM Cloud Pak for Data[27]. As we have outlined earlier, AI on any cloud should be an integrated infrastructure, network, storage, and server (including specialized servers) platform that can be deployed in a set of pre-defined balanced configurations on a variety of cloud platforms, such as IBM Cloud, Red Hat OpenShift, AWS, and so on. It should even include AI applications, accelerators, and assets. This is exactly what IBM Cloud Pak for Data is addressing, as depicted in Figure 4-5.

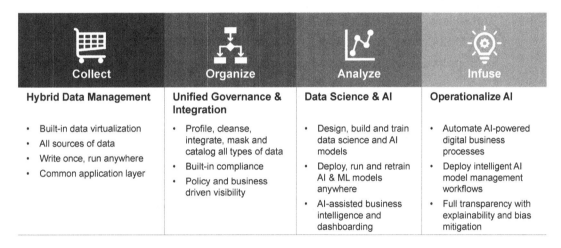

**Figure 4-5.**  *Key Capabilities of IBM Cloud Pak for Data*

# High-Level Description

IBM Cloud Pak for Data is an integrated collection of data and analytics microservices built on a cloud-native architecture, which enables users to collect, organize, and analyze data with unprecedented simplicity and agility, within a governed environment. The software components are pre-configured and require no assembly. It helps organizations to operationalize AI and to accelerate inference to infuse analytical insight

---

[27]See [13] for more details on the IBM Cloud Pak for Data.

into AI applications. In essence, IBM Cloud Pak for Data[28] is a multicloud data and AI platform delivering an AI information architecture with AI capabilities out of the box. It leverages containers and Kubernetes as the foundation. This enables it to run on a private cloud or on a public cloud platform, such as IBM Cloud, Red Hat OpenShift, AWS, Microsoft Azure, and Google Cloud Platform.

The following is a high-level description of key capabilities of IBM Cloud Pak for Data[29]:

- **Data virtualization**: Allows users to discover, access, govern, and analyze remote data without physically moving it. It includes SQL query pushdown optimizations.

- **Databases**: Provides support for integrated data bases, for instance, IBM Db2 Event Store, Db2 Warehouse SMP and MPP deployment options as a single package, Mongo DB, and so on.

- **Multi-tenancy support**: Users and organizations can share the same cluster, yet operate independently in their own dedicated Cloud Pak for Data instances. Each instance has its own isolated users, data, quotas, namespace, and ports[30].

- **Terraform support**: using Terraform to automatically install a stand-alone version of IBM Cloud Pak for Data on AWS or Microsoft Azure, bypassing manual installation steps, including network, firewall, and registry.

- **Industry accelerators**: Three accelerators consisting of pre-configured data attributes, data model, and notebooks for the wealth management space – dynamic segmentation, client attrition, and life event and financial event prediction.

---

[28]See [14] for more detailed information on the content, terms and conditions, prices, and so on on IBM Cloud Pak for Data V2.1.

[29]See [15] for a current description of the latest IBM Cloud Pak for Data capabilities, functions, and features.

[30]See [16] for more information on multi-tenancy of IBM Cloud Pak for Data.

# IBM and Third-Party Add-ons

IBM Cloud Pak for Data V2.1 includes the following six IBM add-ons:

1. **IBM Data Stage Edition for Cloud Pak for Data**: Expands base integration capabilities, providing additional connectors and transform stages, ability to schedule jobs, and Git integration for versioning, backups, and recovery of ETL jobs.

2. **IBM Watson Knowledge Catalog Professional**: Infuses governance and enables self-service for finding, accessing, and preparing data in essence ensuring business-ready data is available to data citizens.

3. **IBM Watson OpenScale for IBM Cloud Pak for Data**: Provides visibility into how AI is built, used, and provides visibility and explainability into AI outcomes. More on this (at a conceptual level, not at a platform level) will be discussed in Chapter 6, "*The Operationalization of AI*," and in Chapter 8, "*AI and Governance*." It can, for instance, detect bias in ML models caused by unexpected correlations from new data sources.

4. **IBM Watson Applications and APIs:**

   a. **Watson Assistant**: Offering for building conversational interfaces into any application, device, or channel

   b. **Watson Discovery**: Makes it easy to build AI solutions that find relevant answers in complex, disparate data with speed and accuracy

   c. **Watson API Kit**: Natural language understanding, Watson Knowledge Studio, Speech to Text, and Text to Speech

In addition, IBM Cloud Pak for Data V2.1 includes the following six third-party add-ons from PostgreSQL, Knowis, WAND, NetApp, Prolifics, and Lightbend. Wand, for instance, provides pre-built taxonomies and ontologies to accelerate building a business glossary.

# Data Virtualization

As we have described previously, the data virtualization engine of the IBM Cloud Pak for Data[31] allows users to discover, access, govern, and analyze remote data without physically moving it. It consolidates data that is available in multiple data sets into a single table, thus virtualizing the data by joining the tables. There is no need for ETL processes and duplicating data storage. Data is not copied; it exists only at the source.

Figure 4-6 is a conceptual depiction of the data virtualization in IBM Cloud Pak for Data. IBM Data Virtualization Manager for z/OS can serve as a data provider for the data virtualization engine in IBM Cloud Pak for Data. This allows, for instance, Z-data to be accessible through a Jupyter Notebook running on IBM Cloud Pak for Data and non-Z-data from any data source under the control of IBM Cloud Pak for Data to be accessible through a Jupyter Notebook running on Watson ML for z/OS.

Thus, Z-data and non-Z-data can be virtualized and made easily accessible by any application running on IBM Cloud Pak for Data.

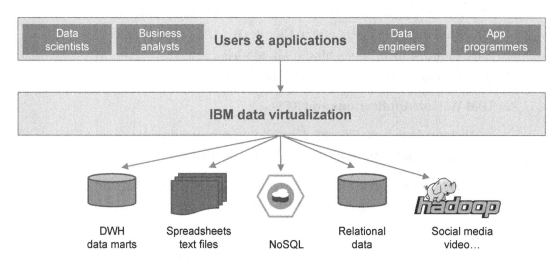

***Figure 4-6.*** *Data Virtualization in IBM Cloud Pak for Data*

Our AI information architecture is mainly represented in the data virtualization methods through the following ABBs in the source data access layer: connectors, virtualization, federation, and APIs.

---

[31]See [17] for more information on data virtualization in IBM Cloud Pak for Data.

# Amazon

Amazon Web Services (AWS) delivers a comprehensive and integrated set of services that enables AI with analytics and ML, based on data lake deployments, which include information architecture services to easily access, move (on-prem and in real time), store, govern, and transform data. AWS favors the usage of a data lake as a centralized repository to store all relevant structured and unstructured data at any scale.[32] Contrary to traditional DWH implementations, a data lake provides the advantage to store data in its original format without knowledge about the future data consumption intent. Therefore, no ETL or specific data model (e.g., a DWH star schema) needs to be implemented upfront. This provides maximum flexibility for business applications.

However, it often comes with high cost and risk associated with the movement of potentially rather large volume of data. In order to address the data movement challenges, AWS provides multiple ways to move data from client data stores to AWS, such as AWS Direct Connect and AWS Snowball and AWS Snowmobile to move petabytes to exabytes of data to AWS. Data can also be moved in real time *from new sources such as websites, mobile apps, and Internet-connected devices*. Amazon Kinesis Data Firehose, Amazon Kinesis Video Streams, or AWS IoT Core can be used to capture and load streaming data to your AWS data lake.

The AWS analytics and ML portfolio provide a rich set of services that correspond to our "*analytics and AI layer*" and "*deployment and operationalization layer*" of our core ABB as depicted in Figure 4-2 AI information architecture overview diagram.

The following list is a short description of the key AWS analytics services:

- **Interactive analytics**: Amazon Athena provides interactive analysis using standard SQL language.

- **Big data processing**: Amazon Elastic MapReduce (EMR) is a cloud-native big data platform, based on open source tools.

- **Data warehousing**: Amazon Redshift is a cloud-based DWH platform for complex, analytical queries against large data volume.

- **Real-time analytics**: Based on Amazon Kinesis, AWS enables you to ingest, buffer, and process streaming data in real time.

---

[32]See [18] for more information on data lakes and analytics on AWS.

- **Operational analytics**: With Amazon Elasticsearch Service, AWS allows operational analytics, providing support for Kibana, and so on.

- **Dashboards and visualizations**: Through Amazon QuickSight, AWS provides services to build visualizations and rich dashboards.

The following list is a short description of the key AWS ML services and tools:

- **Frameworks and interfaces**: AWS provides a number of ML and DL frameworks, such as Apache MXNet, TensorFlow, PyTorch, and others.

- **Platform services**: Amazon SageMaker[33] is an ML platform service for the entire ML workflow, launched in 2017.

- **Application services**: AWS provides AI solution-oriented APIs, which are application services for developers.

# Microsoft

The Microsoft Azure cloud computing platform can be used for building, testing, deploying, and managing applications and services through Microsoft-managed data centers. It includes services that relate to some of our core ABBs of the AI information architecture, as depicted in Figure 4-2 AI information architecture overview diagram. This includes analytics and integration services and the following three pillars of Azure AI services[34]:

1. **Machine learning**: The ML service is a cloud-based service that can be used to build, train, deploy, and manage your AI and ML models. The service includes automated ML and autoscaling functionality, which identifies the best ML algorithms and makes it easy configuring hyperparameters much faster. It includes open source frameworks, languages, and standards such as PyTorch, TensorFlow, Python, R, scikit-learn, ONNX, and so on and enables users to manage and track ML models after deployment.

---

[33]We come back to Amazon SageMaker in Chapter 8, *"AI and Governance."*
[34]See [19] for more information on Microsoft Azure AI.

2. **Knowledge mining**: This AI service can be used to analyze data and search for valuable insights and trends from all your content on a huge scale. It uses Azure Cognitive Search with built-in capabilities to discover patterns and relationships in your content and to understand sentiment. The content can have any format, such as emails, text files, documents, PDFs, images, scanned forms, and so on.

3. **AI apps and agents**: These are a set of Azure Cognitive Services and a Bot Service, which are *pre-trained, ready-to-use algorithms that enable apps, bots, and websites to see, hear, speak, understand, and interpret user needs in a way that feels natural and human.* You can customize these models with your own data and deploy them anywhere. Azure Bot Service allows you to create bots quickly from out-of-the-box templates, manage them, and launch them across a variety of platforms. The Azure Cognitive Services are related to vision, speech, language, decision, and search.

Microsoft Azure also includes advanced security, governance, and control services to protect your assets, restrict access, or to apply Azure security policies.

# Google

Google provides a suite of public cloud services and IaaS under the Google Cloud Platform[35] (GCP). Google clearly stands out with its innovations and open source contributions, especially for container technologies and DL and ML frameworks. Kubernetes, for instance, has evolved from an internal container orchestration effort at Google in the 2003–2004 time frame. TensorFlow, a popular DL framework, has been invented by Google. Recently, Google started an open source project called Kubeflow that tries to bring together ML and DL with Kubernetes, enabling, for instance, to train and deploy TensorFlow DL models in Kubernetes containers. Most of the ABBs of our AI information architecture can be mapped to the GCP services and products, for instance, the compute products, the data analytics products, the migration products for data transfer, and so forth.

---

[35]See [20] for more information on the Google Cloud Platform (GCP).

Google offers a set of AI and ML products as part of the GCP scope, such as the AI Hub and the AI building blocks. In 2019, Google has launched its Google AI Platform[36], which is a comprehensive platform to build *AI applications that run on the GCP and on-premises*. The Google AI Platform is an end-to-end DL and ML platform as a service (PaaS) targeting DL and ML developers, data scientists, and AI infrastructure engineers.

The following list describes the key components of the Google AI Platform:

1. **AI Platform Notebooks**: Enables developers to create and manage virtual machine (VM) instances that are pre-packaged with JupyterLab, the latest web-based interface for Jupyter

2. **Deep Learning VM Image**: Makes it simple to instantiate a VM image containing the most popular DL and ML frameworks on a Google Compute Engine instance

3. **Deep Learning Containers** (beta): Builds your DL project quickly with a portable and consistent environment for developing, testing, and deploying your AI applications

4. **Data Labeling Service** (beta): Lets you request human labeling for a collection of data that you plan to use to train a custom ML model

5. **AI Platform Training**: Allows you to train models using a wide range of different customization options

6. **AI Platform Predictions**: Allows you to serve predictions based on a trained model, whether or not the model was trained on AI Platform

7. **Continuous Evaluation** (beta): Continuously compares and evaluates ML models' predictions to provide continuous feedback for required adjustments

8. **What-If Tool**: For investigation of model performance for various features, optimization options, and different data points

---

[36]See [21] for more information on the Google AI Platform.

9. **Cloud TPU**: Using a set of HW accelerators designed by Google specifically for TensorFlow model processing

10. **Kubeflow**: Using the ML toolkit for deployments of ML workflows on Kubernetes

# Sample Scenarios

In this section, we describe a number of sample scenarios that are related to the core ABBs of the AI information architecture. The term *scenario* is usually used in a rather fuzzy and ambiguous way. Ideally, this term should be used in relationship to a broader set of topics, such as the key actions of the scenario, the required data sources, the data flow, required technical capabilities, derived outcome, monetization aspects, and business value.

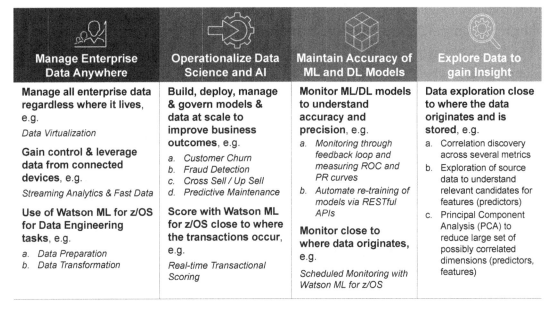

| Manage Enterprise Data Anywhere | Operationalize Data Science and AI | Maintain Accuracy of ML and DL Models | Explore Data to gain Insight |
|---|---|---|---|
| Manage all enterprise data regardless where it lives, e.g. *Data Virtualization* | Build, deploy, manage & govern models & data at scale to improve business outcomes, e.g. <br> a. *Customer Churn* <br> b. *Fraud Detection* <br> c. *Cross Sell / Up Sell* <br> d. *Predictive Maintenance* | Monitor ML/DL models to understand accuracy and precision, e.g. <br> a. *Monitoring through feedback loop and measuring ROC and PR curves* <br> b. *Automate re-training of models via RESTful APIs* | Data exploration close to where the data originates and is stored, e.g. <br> a. Correlation discovery across several metrics <br> b. Exploration of source data to understand relevant candidates for features (predictors) |
| Gain control & leverage data from connected devices, e.g. *Streaming Analytics & Fast Data* | | | |
| Use of Watson ML for z/OS for Data Engineering tasks, e.g. <br> a. *Data Preparation* <br> b. *Data Transformation* | Score with Watson ML for z/OS close to where the transactions occur, e.g. *Real-time Transactional Scoring* | Monitor close to where data originates, e.g. *Scheduled Monitoring with Watson ML for z/OS* | c. Principal Component Analysis (PCA) to reduce large set of possibly correlated dimensions (predictors, features) |

***Figure 4-7.*** *AI Information Architecture-Related AI Scenarios*

As we limit our description to the technical aspects of the AI information architecture, our view is rather narrow and focused on required technical capabilities and the mapping of these scenarios to key ABBs.

As you see in Figure 4-7, we elaborate on the following four sample scenarios. All of them can be further broken down into smaller subsets.

1. Manage enterprise data anywhere.

2. Operationalize data science and AI.

3. Maintain accuracy of ML and DL models.

4. Explore data to gain insight.

# Manage Enterprise Data Anywhere

To avoid data tourism, enterprise data needs to be preferably managed close to its originating systems or where it is finally stored. Moving data around the enterprise, especially without clarity on its usage intent, is not particularly smart. Thus, this scenario manages all enterprise data regardless of where it lives. This requires, for instance, data virtualization and federation capabilities. It may also require fast and often even real-time provisioning of data to consuming systems.

Since the majority of enterprise data for large organizations is on IBM Z (mainframe systems), data engineering tasks, such as data preparation and transformation, need to be done on the IBM Z platform.

The majority of the ABBs from the source data access layer (e.g., connectors, virtualization, federation, etc.) and the data preparation and quality layer (e.g., aggregation, transformation, synchronization, etc.) from our AI information architecture are relevant for this use case. The specific selection and application of ABBs depends on the orientation of the use case.

# Operationalizing Data Science and AI

Chapter 6, *"The Operationalization of AI,"* discusses the operationalization and deployment aspects of data science and AI in a more detailed way. Thus, we only specify the relevant ABBs for this capability or requirement, which are categorized in the deployment and operationalization layer of our AI information architecture, namely, provisioning, inference and model scoring, and model management. Model management also includes model portability, which allows to develop, for instance, an ML model anywhere and import and deploy the model on other platforms running different tools.

To govern ML and DL models requires a number of ABBs from our information governance and information catalog layer,[37] such as search and discovery, lineage, rules and policies, and so on.

## Maintain Accuracy of DL and ML Models

As we have discussed, ML model accuracy and precision can deteriorate over time. To monitor ML model accuracy and precision through a feedback loop and measuring ROC[38] and PR[39] curves and to initiate automatic retraining of models via RESTful APIs are the essential characteristics of this scenario. It is essential to monitor the accuracy and precision close to where the data originates. If the data originates on a different system or platform compared to the one that is used for model development, training, and validation/test, the model portability capability is essential.

The following are the key ABBs that are relevant for this scenario: model management, model monitoring and feedback, retraining, and provisioning.

## Explore Data to Gain Insight

This scenario is an essential one, typically performed iteratively and prior to developing regression or segmentation models. In addition to, for instance, the principal component analysis (PCA) to reduce a large set of possibly correlated dimensions (predictors, features) to a (possibly even smaller) set of uncorrelated dimensions, the data exploration is geared toward gaining a better understanding of relevant candidates for model features (predictors) or to perform correlation discovery across possibly even hundreds of metrics.

---

[37]AI governance aspects will be discussed in more detail in Chapter 8, "*AI and Governance*".

[38]ROC stands for receiver operating characteristic; the ROC curve is a performance measurement for classification problems at various thresholds settings.

[39]PR stands for precision-recall; the PR curve is a plot of the precision and the recall for different thresholds, much like the ROC curve.

In regard to our AI information architecture, the exploration ABB from the analytics and AI layer is applicable. In addition, the information governance and information catalog layer contains a number of ABBs that are relevant as well, such as search and discovery to identify relevant data artifacts and provenance to, for instance, understand the trustworthiness of data sources – to just name a few.

# Key Takeaways

We conclude this chapter with a few key takeaways, summarized in Table 4-2 that are strictly derived from the scope of the AI information architecture. Although this may constitute a small aspect of the overall set of AI needs, the importance and relevance of the AI information architecture as part of any AI endeavor should have become obvious and appreciated by the reader.

***Table 4-2.*** *Key Takeaways*

| # | Key Takeaway | High-Level Description |
|---|---|---|
| 1 | Take a holistic approach to AI | Any AI endeavor requires more than downloading a few freely available open source ML or DL libraries and runtime engines. Take a holistic, enterprise-wide approach and plan for collaboration, integration, specialized engines, open source, enterprise-scale offerings, and others |
| 2 | Include information architecture aspects | An AI journey, regardless of its chosen entry points and initial requirements or use cases, should include key aspects of the AI information architecture. Anticipate future demand and importance of the AI information architecture |
| 3 | Anticipate information governance needs | Information governance and information catalog aspects should be taken into consideration from the beginning. This is more than adherence and conformance to regulatory compliance requirements; it includes aspects, such as MDM, rules policies, provenance, and others |
| 4 | Understand the entire ML workflow | To develop, train, and validate/test, for instance, an ML model is only a relatively small aspect of the entire ML workflow. Understand and take into consideration other key aspects of the ML workflow, such as monitoring accuracy and precision of models, retraining, and so on |

*(continued)*

***Table 4-2.*** (*continued*)

| # | Key Takeaway | High-Level Description |
|---|---|---|
| 5 | Optimize for operationalization | Successful AI deployments require careful planning and optimization for operationalization of AI artifacts (e.g., ML or DL models), including efficient provisioning of transactional and other data for inference and possible real-time scoring scenarios |
| 6 | Consider data gravity aspects | Avoid data tourism wherever possible by leveraging capabilities and ABBs from the AI information architecture, such as data virtualization and federation and data exploration and preparation close to where the data originates |
| 7 | Realize your AI on any cloud needs | Revisit key imperatives to deploy AI on any cloud (e.g., private, public, or hybrid), such as the need to integrate infrastructure, storage, and servers (including special-purpose engines, e.g., FPGAs, GPUs, etc.), or the expandability to include AI applications, models, and other AI artifacts |

# References

[1]  Rob Thomas, IBM, *Think Blog, AI Watson Anywhere,* www.ibm. com/blogs/think/2019/02/enabling-watson-anywhere/ (accessed May 13, 2019).

[2]  Godinez, S., Hechler, E., Koenig, K., Lockwood, S., Oberhofer, M., Schoeck, M. *The Art of Enterprise Information Architecture: A Systems-Based Approach for Unlocking Business Insight.* ISBN-13: 978-0137035717, Pearson Education (IBM Press), 2010.

[3]  Resmini, A. (Editor), *Reframing Information Architecture,* ISBN-13: 978-3319064918, Springer, 2014.

[4]  The Open Group. *The TOGAF Standard – Version 9.2.* www. opengroup.org/togaf (accessed May 23, 2019).

[5]  Wikipedia. *Reference architecture,* https://en.wikipedia.org/ wiki/Reference_architecture (accessed May 23, 2019).

[6]   The Open Group. Introduction to the ADM, `http://pubs.opengroup.org/architecture/togaf8-doc/arch/chap03.html` (accessed May 25, 2019).

[7]   Dreibelbis, A., Hechler, E., Milman, I., Oberhofer, M., Van Run, P., Wolfson, D. *Enterprise Master Data Management – An SOA Approach to Managing Core Information.* ISBN-13: 978-0132366250, Pearson Education (IBM Press), 2008.

[8]   ONNX. *Open Neural Network Exchange (ONNX) Format.* `https://onnx.ai/` (accessed May 26, 2019).

[9]   Wikipedia. *Predictive Model Markup Language (PMML).* `https://en.wikipedia.org/wiki/Predictive_Model_Markup_Language` (accessed May 26, 2019).

[10]  Nohria, R., Santos, G. IBM Power System AC922 – Technical Overview and Introduction, REDP-5494-00, IBM Redbooks, 2018.

[11]  Mohri, M., Rostamizadeh, A., Talwalkar, A. *Foundations of Machine Learning.* ISBN-13: 978-0262039406, The MIT Press, 2nd edition, 2018.

[12]  Sloan, J., Zawacki, A. *The IBM analytics and machine learning advantage – Optimize the valuable data behind your firewall.* IBM Analytics Thought Leadership White Paper, `www.ibm.com/downloads/cas/NXLXQ8NJ` (accessed May 29, 2019).

[13]  IBM. *IBM Cloud Pak for Data System.* `www.ibm.com/products/cloud-pak-for-data/system` (accessed May 31, 2019).

[14]  IBM. IBM Cloud Private for Data V2.1 delivers an information architecture for developing AI applications with advanced data governance and transformation. `www.ibm.com/downloads/cas/US-ENUS219-305-CA/name/ENUS219-305.PDF` (accessed June 2, 2019).

[15]  IBM. IBM Knowledge Center – Overview of IBM Cloud Pak for Data. `www.ibm.com/support/knowledgecenter/SSQNUZ_2.1.0/com.ibm.icpdata.doc/zen/overview/overview.html` (accessed June 2, 2019).

[16]    Mascarenhas, C., MacKenzie, B., Srinivasan, S. *Multi-Tenancy with ICP for Data.* https://ibm.ent.box.com/s/niud78bitlzgm39vnql40i76u8b2vu46 (accessed June 2, 2019).

[17]    IBM. *Eliminate data silos: Query many systems as one – Data virtualization in IBM Cloud Private for Data.* www.ibm.com/downloads/cas/97AJPYNN (accessed June 4, 2019).

[18]    Amazon. AWS. Data Lakes and Analytics on AWS. https://aws.amazon.com/big-data/datalakes-and-analytics/?nc=sn&loc=1 (accessed February 11, 2020).

[19]    Microsoft. Microsoft Azure. Azure AI. https://azure.microsoft.com/en-us/overview/ai-platform/ (accessed February 11, 2020).

[20]    Google. Google Cloud. Google Cloud Platform. https://cloud.google.com/docs/ (accessed February 12, 2020).

[21]    Google. Google Cloud. AI Platform. https://cloud.google.com/ai-platform/ (accessed February 12, 2020).

# CHAPTER 5

# From Data to Predictions to Optimal Actions

The concept of optimizing decisions based on predictions considering additional data and constraints introduced in Chapter 1, "*AI Introduction*," is often critical to solve real business problems. Decision optimization (DO) takes predictive insight one step further and guarantees that an optimal combination of business-relevant actions can be taken based on predictive insight with relevant context.

In this chapter, we explore this area in more depth and give practical examples of how ML and DO can be combined to get from data to predictions to optimal decisions and resulting actions.

## Use Case: A Marketing Campaign

At the beginning of a new year, a bank wants to run a targeted marketing campaign in order to maximize revenue across its banking products and customer base, while avoiding sending customers too many messages.

The bank owns comprehensive customer profile information, data about customers' deposits with the bank, income, age, household size, and data about what investments and products of the bank customers already have as of the past year.

The bank's data steward team curates the data that is deemed necessary for the marketing campaign project and makes it available to a team of data scientists along with responses to prior marketing campaigns, so that they can create ML models to be used in order to optimize revenue and profit. For this campaign, targeting the most promising customers with the most suitable products and services is of essence.

© Eberhard Hechler, Martin Oberhofer, Thomas Schaeck 2020
E. Hechler et al., *Deploying AI in the Enterprise*, https://doi.org/10.1007/978-1-4842-6206-1_5

# Naïve Solution: ML 101

The team of data scientists builds an ML model for predicting which customers are likely interested in what products or services in the new year.

The predictive model is trained using input data composed of relevant features derived from customer profile information, customer's cash deposits, the products and services they already have in their portfolio, and the products they bought in the past year.

The data scientists provide the ML model to the IT department, with IT team members to deploy and to run a batch scoring job, computing daily for all customers of the bank which products and services they are most likely to buy.

The result is a table in a database with a row for each of the tens of thousands of customers of the bank, including the predicted product each customer individually would be most likely to buy.

These predictions including the probability of buying could be used to send each customer a marketing message, offering them the product that they are most likely to buy; however, in doing so, the bank might run into difficulties:

- **Budget constraints**: The budget for the marketing campaign might be exceeded. The campaign may not be sufficiently targeted, meaning that too many clients may receive an offer that they simply reject. As a result, too much money would be spent without getting an adequate return for it.

- **Product and service availability**: The bank might run out of some products it only has limited financial backing for or may face constraint in being able to provide a time- and resource-intensive service to only a handful of selected clients. If too many clients are offered to buy such products, the bank may not be able to fulfill what they offered with defined quality and services standards.

- **Risks, limitations, and costs**: Product and service offers with associated risks, limitations, and cost require careful additional consideration to avoid negative effects on profitability or risk taken on by the bank.

- **Risk of customer negativity**: Product offers that are perceived as "creepy," giving the customer a sense that the bank knows and uses too much data about them, can cause anger and result in negative perception of the bank.

In practice, a brute-force approach to send every customer an offer for the product predicted to be what they are most likely to buy, disconnected from context and real-world constraints, would be suboptimal. It could cause high cost, risk, and potentially customer dissatisfaction if making too many offers the bank could in the end not all fulfill.

## Refined Solution: ML plus DO

The preceding naïve solution falls short because it makes decisions and takes actions without considering context and constraints. To act on each prediction could make sense if the number of predictions would be much smaller than available resources. But as soon as predictions would lead to a nontrivial number of decisions and resulting actions, it becomes important to consider the whole set of possible decisions in the context of related data and constraints. Driving to the most optimal relevant and context-aware decisions is what decision optimization (DO) solves.

With DO, data scientists or optimization experts can define an *optimization problem* consisting of a set of constraints that need to be honored, objective(s) to be optimized, and data to be considered in solving the problem. The problem defined this way is then solved using a DO engine, generating an *optimized* set of decisions as a result.

Applying DO to the bank's marketing campaign project ensures that within a given marketing budget and product quantities, the optimal set of product offers is made to the right customers based on the predictions from the predictive ML model, optimizing revenue and preventing offers that the bank might not be able to fulfill.

## Example: ML plus DO

In this section, we show you an example how ML and DO can be used together in IBM Watson Studio[1], by creating a predictive ML model and a prescriptive DO model in a project, deploying both artifacts and making them available for use by a process that can invoke the predictive ML model followed by solving the prescriptive DO model[2].

---

[1]See [1] for more information on IBM Watson Studio.

[2]We are using IBM Watson Studio for this example; however, similar tools from a variety of different vendors could be used as well.

# Create a Project

A project owner creates a project and can add various personas, such as data scientists, subject matter experts, and optimization experts as needed to work on the project as a team. Only project members can access the secure environment provided by the project.

As you can see in Figure 5-1, within the scope of a project, these different personas efficiently collaborate with each other to perform various tasks, such as connecting to data sets and refining them. Furthermore, various notebooks can be used for the data to be explored, analyzed, and visualized. Under the umbrella of the defined project, ML and DL models as well as DO models can be developed, trained, validated, deployed, and finally tested.

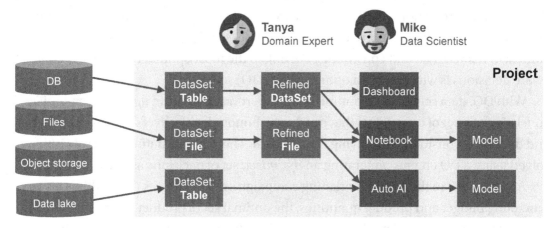

***Figure 5-1.***  *Creating a Data Science Project*

In the following sections, we describe these various tasks in more detail and visualize some of these steps using IBM Watson Studio.

# Connect Data

Project members can connect to data in databases, object stores, or other data sources via connections and reference or copy subsets of relevant data as data assets in the project. If the original data contains sensitive information that is not intended to be used in the project, the person who adds it to the project can omit certain columns when adding the data to the project.

In this case, a data steward was added to the project to make the required data available for use in the project. The data steward selects the data that data scientists, data engineers, and other personas are allowed to use for the project and adds it to the project as a database table or CSV file. Data from various data source systems can be provisioned.

The objective is to make available all data that is relevant for this particular business solution. This sounds like a rather trivial task, which is often grounded on the assumption that relevant data is simply available and easily accessible. Nevertheless, making relevant data available for data engineers and data scientists can be a challenging endeavor.

| customer...<br>String | age<br>String | age_youngest_...<br>String | debt_equ...<br>String | gender<br>String | bad_pay...<br>String | gold_card<br>String | pension_...<br>String |
|---|---|---|---|---|---|---|---|
| 15 | 45 | 12 | 45 | 0 | 0 | 0 | 0 |
| 16 | 43 | 12 | 43 | 0 | 0 | 0 | 0 |
| 30 | 23 | 0 | 23 | 0 | 0 | 0 | 0 |
| 42 | 35 | 8 | 35 | 1 | 0 | 0 | 0 |
| 52 | 43 | 12 | 43 | 1 | 0 | 0 | 0 |
| 57 | 51 | 19 | 51 | 1 | 0 | 0 | 0 |
| 74 | 31 | 0 | 31 | 1 | 0 | 0 | 0 |
| 74 | 31 | 0 | 31 | 1 | 0 | 0 | 0 |
| 89 | 46 | 11 | 46 | 1 | 0 | 0 | 0 |
| 90 | 70 | 38 | 70 | 0 | 0 | 0 | 0 |
| 95 | 39 | 11 | 39 | 1 | 0 | 0 | 0 |
| 105 | 31 | 0 | 31 | 0 | 0 | 0 | 0 |
| 106 | 36 | 7 | 36 | 0 | 0 | 0 | 0 |

**Figure 5-2.** *Data Sets Associated with a Project*

The data sets include relevant columns from the bank's customer records, as illustrated in Figure 5-2 that shows the data set in the data preview in the project.

# Refine, Visualize, Analyze Data

Project members can explore and refine data as needed in order to achieve the required level of data quality and representative data distribution. Data engineers are typically in the lead to prepare and transform the available data into a format that is suitable for further consumption by data scientists. Nevertheless, the data exploration, refinement, and visualization is typically a rather collaborative undertaking, where data scientists work in concert with data engineers. Business domain experts, such as the marketing campaign experts, need to provide guidance to guarantee the required business outcome. Some data may not be useful to make predictions for the future, but still should be included to enable other insights, for example, branch codes and indication of pre-/post-merger records could enable insights such as how these dimensions correlate with performance.

The data refinement can, for instance, be done using the refinery function in the context of the project. It can also be done via coding to refine the data in notebooks[3] using Python, R, or many other languages.

They can be used to interactively visualize and analyze data to better understand correlation and coherence of various data segments. The overall goal is to explore and refine the data, in order to determine relevant features for the creation and training of models.

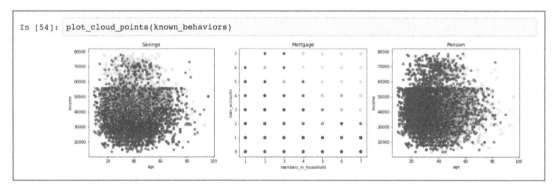

**Figure 5-3.** *Data Visualization in a Notebook*

As you can see in Figure 5-3, in this example, the data scientists use visualizations directly in a notebook to visualize the data regarding customer purchases from the previous year. Many visualization graphics are possible to use, for instance, discovering important correlation between data columns, KPIs, or defined measures.

This allows a data scientist to gain relevant insight of the data segments, which should then be taken into consideration when developing an ML or DL model.

The following is some insight that can be seen from the visualization in Figure 5-3:

- **Customer income**: The greater a customer's income, the more likely it is that they own a savings account. As we have already pointed out in Chapter 1, "*AI Introduction,*" this finding by itself may not be

---

[3]See [2] and [3] for more information on notebooks for data scientists.

sufficient. In other words, it should be correlated with other insights from this particular customer's transactional records or other customer profile information.

- **Customer age**: The older a customer is, the more likely it is that they own a pension account. This finding may be statistically correct and could very well be used in the development of a ML model; however, it needs to be correlated with other insights for a particular client. For instance, a particular client may already have a pension account.

- **Correlation discovery**:[4] There is a correlation between the number of residents in a customer's household, the number of loan accounts held by the customer, and the likelihood a customer buys a mortgage account, as can be seen in the upper right and lower left corners of the mortgage chart in Figure 5-3.

The preceding examples also emphasize nicely that possible bias should be taken into consideration. The insight could, for instance, be biased toward certain demographic measures (e.g., age, gender, marital status, etc.), which may or may not reflect the reality.

# Create and Train Predictive Models

Once the data is in the right shape and well understood, data scientists can create and train predictive machine learning models using integrated tools, for example, using Auto AI, Python Notebooks, or SPSS flows.

We provide some more details of these integrated tools, to give you a better understanding about their capabilities.

## Auto AI

Auto AI provides an easy way to create a set of model pipeline candidates by providing a data set and letting Auto AI[5] perform model selection, feature engineering, hyperparameter optimization, and others for a set of pipeline candidates. Data scientists can then explore various metrics of the resulting models, pick the models they like best, and save them to the project.

---

[4]Correlation discovery is an essential aspect of ML; see [4] for more information on correlation discovery and its applicability in ML.

[5]See [5] for more information on Auto AI.

The resulting models can be evaluated regarding their accuracy and precision by comparing the areas under the ROC and PR curves. This allows a consistent comparison of all models and picking the best model to use.

Auto AI is a great example of how you can get started with AI and ML projects, without necessarily being a data science subject matter expert and skilled in mathematical and statistical methods. It hides some of the complexity that is typically associated with data science tasks.

***Figure 5-4.*** *Auto AI Experiment Pipeline Generation*

Figure 5-4 shows a summary of the Auto AI pipeline generation with various steps, such as selecting the algorithms, performing hyperparameter optimization, and feature engineering tasks.

Some of these steps can be performed iteratively to increase the accuracy and precision of the models. As Auto AI executes the steps, it shows progress to the user. When larger data sets are used, users can also sign off and return to their experiment later.

Figure 5-5 shows an overview of all the selected algorithms, candidate training pipelines created based on these algorithms, and feature transformers used by these pipelines after an Auto AI experiment concluded. Apart from picking the model that best fits your needs and saving it, you can also pick a model pipeline and can save it as a Python Notebook.

This allows you to further improve and customize various tasks (e.g., data preparation, feature engineering, hyperparameter settings) and further optimize the accuracy and precision of your ML models.

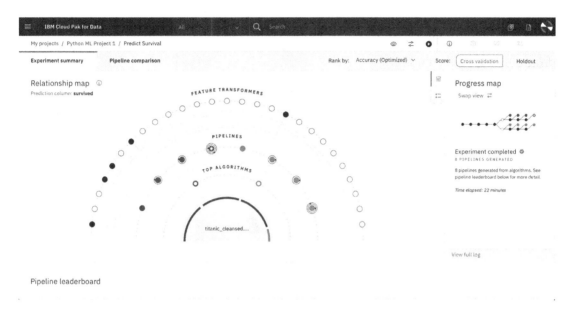

***Figure 5-5.*** *Auto AI Experiment Summary*

You can also optionally interact with Auto AI to specify your own preferences in the automated Auto AI process. The following are some examples:

- Data preparation and advanced data refinery tasks

- Feature engineering, including feature transformations

- Auto AI pipeline optimization

- Hyperparameter optimization (HPO)[6]

- Explainability, debiasing, and fairness

- AI life cycle management to monitor post-deployment performance

---

[6]See [6] for more information on hyperparameter optimization.

## SPSS Flows

SPSS flows allow multiple personas – including business domain experts without coding skills – to create and train models by defining model training flows in a visual editor and running these flows to create, train, and save models to the project.

Figure 5-6 is an illustration of a sample SPSS flow, which can be assembled with no programming skills required.

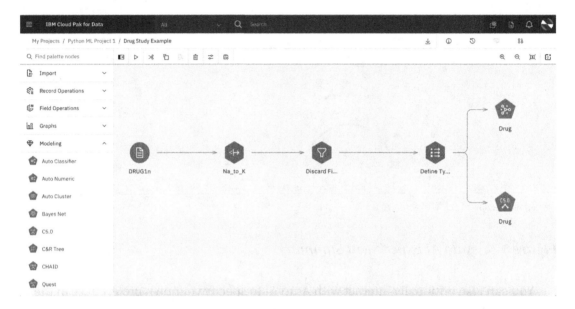

***Figure 5-6.*** *Sample SPSS Flow*

SPSS[7] is an alternative for personas with limited or none programming skills.

## Notebooks

As we have mentioned before, notebooks can be used alternatively by data scientists to train models using Python or many other languages like R or Scala, where the trained models can then be saved to the project using the project API from their code in the notebook. Notebooks require programming skills; however, they allow for greatest flexibility and need to be considered as state-of-the-art technique for any data scientist.

---

[7]See [7] for more information on SPSS.

Python is what we observe to be the most popular programming language among data scientists. Python is also a very popular programming language in general. This has led to a wide range of readily available libraries that can be used from Python code, including a large number of powerful data science, ML, DL, DO, and visualization libraries that make data scientists highly productive.

Figure 5-7 is a snapshot taken while working in a project with a Jupyter Notebook opened in the JupyterLab user interface. It shows the typical combination of displaying code cells and resulting output in a notebook document, which enables literate programming. A notebook is self-documenting; after running, it contains code and resulting insights together with inline text explaining it all. As you can also see in Figure 5-7, JupyterLab can integrate with a Git repository for code management, in this case with a Git repository that was associated with the project.

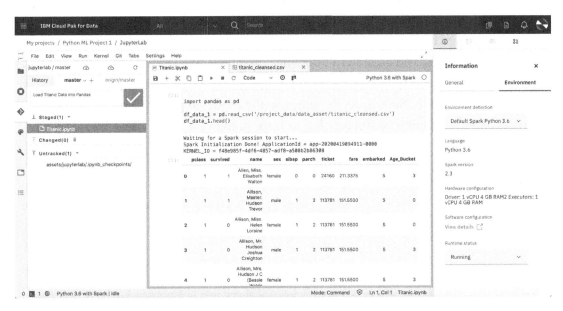

***Figure 5-7.*** *Jupyter Notebook with Python*

In order to run code cells, notebooks need an underlying runtime environment, which in this case is a Python + Spark environment, allowing both single node execution of Python code as well as parallel execution of Python code leveraging the Apache Spark framework.

```
In [18]:   from sklearn import svm
           from sklearn import ensemble

In [19]:   classifiers = []
           for i,p in enumerate(products):
               clf = ensemble.GradientBoostingClassifier()
               clf.fit(X, ys[i])
               classifiers.append(clf)
```

***Figure 5-8.*** *Python Notebook to Train a Model*

An excellent example illustrating the combination of ML and DO was created by Alain Chabrier[8] from IBM, thought leader and expert in DO and in combining it intelligently with ML. As part of his work in this field, he created a notebook using a banking scenario as an example, from which we included excerpts in this chapter. To generate predictions of customer interest in products, the first step is to train a model using data of customer purchases and customer profile data from the past year.

As can be seen in Figure 5-8, *scikit-learn svm* and *ensemble* are imported to be used in the notebook, and a gradient boosting classifier is used as the underlying ML algorithm. The gradient boosting algorithm is a very popular one for regression and classification problems. Subsequently, they use the model to predict which customers will most likely buy which particular offer in the coming year.

```
In [24]:   import warnings
           warnings.filterwarnings('ignore')

In [25]:   predicted = [classifiers[i].predict(to_predict) for i in range(len(products))]
           for i,p in enumerate(products):
               to_predict[p] = predicted[i]
           to_predict["id"] = unknown_behaviors["customer_id"]
```

***Figure 5-9.*** *Using the Model for Scoring*

As you can see in Figure 5-9, we are taking each product (savings, mortgage, and pension) and relate this to the chosen customer characteristics, such as age, income, number of members in a household, and number of accounts.

---

[8]See [8] for more information on DO from Alain Chabrier.

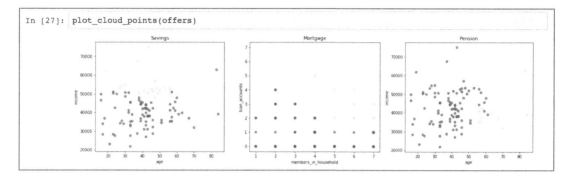

**Figure 5-10.** *Visualization of Predicted Offers*

The predictions for what to offer can be visualized in the same way as previously the data from the past year was displayed. As you can see in Figure 5-10, we receive a predictive outcome that suggests the following for the individual products:

- **Savings**: There is a higher likelihood for customers with greater income above US $ 50.000 p.a. in need of a savings account.

- **Mortgage**: There is a correlation between the number of members in a household and the number of loan accounts. Households with more than five members and a minimum of two loan accounts are more likely to need a mortgage.

- **Pension**: There is a higher likelihood for customers with an age of greater than 56 and an income of US $ 42.00 or lower to need a pension product.

This example is an illustration; for a realistic scenario, additional product and customer characteristics need to be taken into consideration as well.

# Deploy ML Models

Models can be promoted from a project to a *space*, where authorized users can create *model deployments* to serve the models for online or batch scoring.

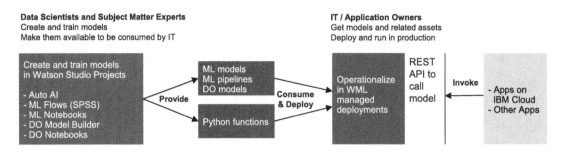

**Figure 5-11.** *Deploying ML Models*

This makes the predictive ML or DL models accessible through public REST APIs. Applications or business processes can invoke the models through these REST APIs to get predictions. Figure 5-11 depicts the deployment of ML or DL models. Once models are deployed, payload input and output logs with model input data and model prediction output can be recorded in a database table, which can be continuously monitored and analyzed for fairness. This allows to automatically detect poor performance, drift, or bias in model scoring and enables taking corrective action if needed.

Subject matter experts create, train, and validate models and make them available to be operationalized within the IT and business environment. The AI artifacts could be ML or DL models and pipelines, DO models, and Python functions to combine models. IT and applications need to integrate these AI artifacts to be called via REST APIs for scoring.

## Create DO Models

As depicted in Figure 5-12, in order to progress from predictions to optimal actions, the team needs to combine ML with DO, so that predictions from an ML model plus other input data can feed into a prescriptive DO model to finally determine optimal actions based on that input.

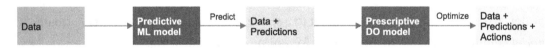

**Figure 5-12.** *Creating DO Models*

An optimization expert or data scientist familiar with optimization creates and tests a DO model in a notebook or in the DO Model Builder, using data from the project and predictions created by the predictive ML or DL model, to solve for optimal decisions and resulting actions. DO in IBM Watson Studio leverages the advanced docplex engine[9] in order to solve optimization problems.

In this example, being familiar with Python, the data science team chooses to create a Python Notebook and then define and test the DO model in Python code in the notebook. While the docplex Python library is already preinstalled in Python environments in IBM Watson Studio, it can also easily be added in any Python Notebook elsewhere using !pip install docplex, providing you with the greatest flexibility. Then a DO model can be created using the following code in a Python Notebook cell:

```
from docplex.mp.model import Model
mdl = Model(name="marketing_campaign")
```

After this step, the decision variables need to be defined, followed by defining the constraints that need to be considered. In this example, the following constraints are relevant:

- Offer only one product per customer.

- Compute the budget and set a maximum on it.

- Compute the number of offers to be made.

- Ensure at least 10% of offers are made via each channel.

```
In [34]:   # At most 1 product is offered to each customer
           mdl.add_constraints( mdl.sum(channelVars[o,p,c] for p in productsR for c in channelsR) <=1
                             for o in offersR)

           # Do not exceed the budget
           mdl.add_constraint( mdl.sum(channelVars[o,p,c]*channels.get_value(index=c, col="cost")
                                                 for o in offersR
                                                 for p in productsR
                                                 for c in channelsR)  <= availableBudget, "budget")

           # At least 10% offers per channel
           for c in channelsR:
               mdl.add_constraint(mdl.sum(channelVars[o,p,c] for p in productsR for o in offersR) >= len(
           offers) // 10)

           mdl.print_information()

           Model: marketing_campaign
            - number of variables: 900
              - binary=900, integer=0, continuous=0
            - number of constraints: 104
              - linear=104
            - parameters: defaults
```

***Figure 5-13.***  *Defining Constraints for a DO Model*

---

[9]See [9] for more information on the docplex engine.

These constraints are defined for the model using `mdl.add_constraint()`, as can be seen in Figure 5-13.

---

**Express the objective**

You want to maximize expected revenue, so you take into account the predicted behavior of each customer for each product.

```
In [35]: obj = 0

         for c in channelsR:
             for p in productsR:
                 product=products[p]
                 coef = channels.get_value(index=c, col="factor") * value_per_product[product]
                 obj += mdl.sum(channelVars[o,p,c] * coef* offers.get_value(index=o, col=product) for o
         in offersR)

         mdl.maximize(obj)
```

***Figure 5-14.*** *Defining the Objective*

Then the objective is defined, in this case to maximize revenue using `mdl.maximize()`, as can be seen in Figure 5-14.

Finally, the docplex engine[10] is run to solve the model using `mdl.solve()`, as can be seen in Figure 5-15.

---

```
In [38]: s = mdl.solve()
         assert s, "No Solution !!!"

In [39]: print(mdl.get_solve_status())
         print(mdl.get_solve_details())

         JobSolveStatus.OPTIMAL_SOLUTION
         status  = integer optimal solution
         time    = 0.0108402 s.
         problem = MILP
         gap     = 0%
```

***Figure 5-15.*** *Running the Docplex Engine*

The output of solving the model is a data frame with the detailed optimized decisions for what offers to make to which customers via which channel.

---

[10]More information on how to use docplex in notebooks can be found here: https://github.com/IBMDecisionOptimization/docplex-examples.

# Deploy DO Models

Like predictive ML models, prescriptive DO models can be deployed to solve optimization problems. This makes the DO models accessible through the Watson ML public REST APIs for access by applications or business processes. Subsequently, with relevant data and ML predictions as input, the docplex engine will solve the problem and generate optimal actions.

This represents a holistic solution in the very specific context of the process or business application.

By deploying both ML and DO models in combination with the same Watson ML service, it becomes easy for applications or processes to call ML models to generate predictions based on relevant data and to then call DO models with data predictions to determine the optimal actions to take.

In this example, the predictions and the optimized decisions are persisted to database tables resulting from batch scoring deployments of the ML and DO models.

# Taking ML and DO Models to Production

As we observed in the first chapter, in order to deploy models for production use, they often need to be added and deployed to completely separate production deployment systems.

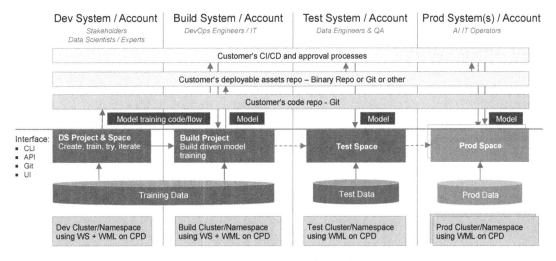

***Figure 5-16.*** *Heterogeneous Development and Production Environment*

To achieve this with Watson Studio, it is possible to, for example, install one OpenShift cluster with IBM Cloud Pak for Data with Watson Studio and Watson Machine Learning as a working environment for data scientists and to install additional clusters with Watson Machine Learning for build, test, and production environments, as depicted in Figure 5-16.

Data scientists can submit their work results to a Git repository or export the assets as a ZIP file, in order to make the ML and DO model assets available to a CI/CD[11] pipeline created and operated typically by a separate IT team. The CI/CD pipeline can then propagate assets to build, test, and eventually production environments, to establish a well-defined process for taking ML and DO model assets to production after reproducible training and testing.

## Embedding AI in Applications and Processes

Having deployed ML and DO models for production use, business processes and applications can now use these models to make predictions and determine optimal decisions to drive optimal automated or AI augmented actions.

In our example, the bank's marketing processes can obtain the resulting optimized decisions from a database table that contains the optimal set of decisions for what offers to make to which customers through which channel in order to achieve the optimal result within the available budget and product availability and other constraints.

## Key Takeaways

We conclude this chapter with a few key takeaways, summarized in Table 5-1.

---

[11]CI/CD stands for continuous integration and continuous delivery.

**Table 5-1.** *Key Takeaways*

| # | Key Takeaway | High-Level Description |
|---|---|---|
| 1 | Getting from data to predictions is often not enough | Predictions alone are typically not actionable; it is key to get from predictions to optimal decisions in order to inform the best actions |
| 2 | Naïve approach for a marketing campaign | For example, train a model to predict which customers are likely to buy what products – but in practice may not be feasible to market to all potential buyers |
| 3 | Smarter approach for a marketing campaign: consider constraints | Take constraints into account, such as budget, availability of product, mailing limitations, and so on and optimize toward well-defined targets |
| 4 | Decision optimization for automated solving of optimization problems | With decision optimization, data scientists or optimization experts can define an *optimization problem* consisting of a set of constraints that need to be honored, objective(s) to be optimized, and data to be considered in solving the problem |
| 5 | IBM Watson Studio is an example of an environment combining ML + DO | Watson Studio allows to create projects, add members, connect and add data, build machine learning models and decision optimization models, and deploy ML and DO models together for use by applications and processes |
| 6 | Decision optimization models can be created via UI or Python | DO designer allows to define models visually; alternatively, DO models can be defined and solved in Jupyter using Python |
| 7 | Use DevOps with CI/CD for taking ML and DO models to production | Ensure all relevant artifacts are managed in a trusted code and asset repository (e.g., Git) and can be deployed to test and production systems via automated CI/CD pipelines in a reproducible fashion |
| 8 | Achieving HA and DR for model deployments | Deploy models to at least two independent sites with load balancing of requests and with failover if one site fails to serve business critical applications and processes |

# References

[1]   IBM, *IBM Watson Studio*, www.ibm.com/cloud/watson-studio (accessed April 27, 2020).

[2]   Galea, A. *Beginning Data Science with Python and Jupyter: Use powerful tools to unlock actionable insights from data.* ISBN-13: 978-1789532029, Packt Publishing, 2018.

[3]   Nelli, F. *Python Data Analytics: With Pandas, NumPy, and Matplotlib.* 2nd ed. Edition. ISBN-13: 978-1484239124, Apress, 2018.

[4]   Mirkin, B. *Core Data Analysis: Summarization, Correlation, and Visualization* (Undergraduate Topics in Computer Science) 2nd ed. 2019 Edition. ISBN-13: 978-3030002701, Springer, 2019.

[5]   Malaika, S., Wang, D. IBM. Artificial Intelligence. *AutoAI: Humans and machines better together*, https://developer.ibm.com/technologies/artificial-intelligence/articles/autoai-humans-and-machines-better-together/ (accessed April 27, 2020).

[6]   Naya, G. Towards Data Science. *Available hyperparameter-optimization techniques*, https://towardsdatascience.com/available-hyperparameter-optimization-techniques-dc60fb836264 (accessed April 27, 2020).

[7]   IBM. *IBM SPSS software*, www.ibm.com/analytics/spss-statistics-software (accessed April 27, 2020).

[8]   Chabrier, A. *Decision Optimization for Data Science Experience: Why?* https://developer.ibm.com/docloud/blog/2018/05/07/do4dsx-why/ (accessed April 27, 2020).

[9]   IBM, *The IBM Decision Optimization CPLEX Modeling for Python*, https://pypi.org/project/docplex/ (accessed April 20, 2020).

# CHAPTER 6

# The Operationalization of AI

Developing AI solutions, including the training and deployment of ML/DL models, remains an important and often IT resource-intensive task. The integration of AI artifacts, such as ML/DL models and data engineering modules, into an existing enterprise IT infrastructure and application landscape represents an additional challenge. The productization or operationalization of AI and the inference of AI-based analytical insight into consuming applications are further explored in this chapter, where we focus on the productization and operationalization of AI specifically in an enterprise context. Furthermore, we shed light on the key challenges of operationalizing AI[1] and describe essential goals for an efficient and sustainable productization of AI solutions, particularly ML and DL models and data engineering artifacts.

## Introduction

Although AI and ML are not new at all, the degree of how insight-driven organizations are embedding ML and DL into their business processes and use cases via software and applications varies greatly. Especially, operationalizing ML models and ensuring real-time scoring still represent a huge challenge for most organizations. It is related to the required efficiency of model deployment, model pipelining and scoring, auditing, and monitoring.

One particular aspect is to operationalize predictive models within transactional applications, without significant overhead, enabling real-time insight at the point of interaction. Real-time transactional scoring or scoring with ultralow latency is a key requirement for quite a number of use cases, for instance, for fraud protection, contextual marketing offers, cybersecurity, and so on.

---

[1]See [1] for more information on the operationalization of AI deployments.

© Eberhard Hechler, Martin Oberhofer, Thomas Schaeck 2020
E. Hechler et al., *Deploying AI in the Enterprise*, https://doi.org/10.1007/978-1-4842-6206-1_6

AI deployments need to be optimized for efficient operationalization of ML and DL in enterprise environments, which require confidence, robustness, and adequate performance. To deliver essential model versioning, auditing, and monitoring as well as high availability, high performance, and low latency is vital for an enterprise-ready ML and DL solution. Versioning allows deployment of different model versions, including scoring of different versions using different scoring services – if needed. Model monitoring and retraining and model adjustments ensure enterprise relevance of models over their lifetime. To ensure accuracy and precision of models, retraining of models should, for instance, be implemented via RESTful APIs.

Another rather important aspect is ML and DL model automation (which we call ML and DL as a service). To enable AI operationalization[2], vendor offerings and tools should enable data engineering and pipelining capabilities, which is comprised of reading and provisioning the required data for the scoring engine; converting the data, for instance, to a Pandas or Spark DataFrame; dropping or adding some columns; performing some transformations or calculations over the columns; and normalizing the data. Furthermore, vendor offerings and tools should, for instance, provide various options for scoring of ML and DL models. Options could, for instance, be based on an online memory-based scoring engine deployment or a CICS-integrated online scoring service for enterprises with a traditional mainframe infrastructure. Today's enterprises may use several ML and DL model types, such as Apache Spark models, scikit-learn models, and PMML or ONNX models, which also require different scoring engines.

The necessary data engineering tasks including the data access, data pipelining, and transformation tasks in preparation for model scoring require an optimized IT infrastructure and enterprise readiness that most enterprises are still struggling with today.

# Challenges of AI Operationalization

There are a number challenges associated with the AI operationalization and productization. The following are key areas or domains, which we use to categorize these AI operationalization challenges[3]. These following six domains are distinct, but still related to each other.

---

[2]A subset of the scope of AI operationalization as used in this book is often referred to as *inference* or *inferencing*.

[3]See [2] for more information on requirements and challenges of AI operationalization.

1. **Data engineering**: As input to the scoring service, data engineering and pipelining is required, which includes the preparation, transformation, and pipelining of new data records, sensor device input, and others into features that can be consumed and processed by ML and DL models. This represents a huge challenge, since source data may have to be assembled with ultralow latency from a heterogeneous system and application landscape. Some level of data pipelining may be integrated in so-called AI pipeline models.

2. **Infrastructure transparency**: ML and DL models may be developed, trained, and validated on one particular IT platform (e.g., a Hadoop-based data lake), using a specific set of AI open source or vendor offerings, whereas the deployment and operationalization needs to be done on a different platform (e.g., a traditional mainframe system) with other offerings and tools, which requires IT infrastructure transparency and interoperability of open source tools and vendor offerings.

3. **Model scoring**: Online scoring services have to be *integrated* into the existing transactional or application landscape, which requires flexibility and high performance, as can, for instance, be provided via microservices and RESTful APIs or by integrating scoring services directly into a database management system (DBMS) or into the existing transaction management systems. As an example, IBM Watson ML for z/OS scoring services can be configured in a CICS region.[4]

4. **Insight inference**: After the scoring of ML/DL models, inference of the analytical insight needs to take place. The outcome of the model, for instance, a score of a classification model or a set of clusters of an unsupervised clustering model, needs to be infused into corresponding applications. This requires clarity and interpretability of ML/DL model outcome.

---

[4]See [2] for more information on configuring IBM Watson ML for z/OS scoring services in a CICS rcgion.

5. **Model monitoring**: Once deployed and operationalized, ML/DL model monitoring is needed to detect, understand, and measure, for instance, bias and deviations from the original accuracy and precision[5] of those models. Depending on the requirement of the use case and application, this needs to be done continuously or at least periodically.

6. **Model adaptations**: The insight, which is derived from ML/DL model monitoring, may suggest adjustments of the AI application or of some models, either resulting into retraining of the existing models with new labeled training and validation data or even model adjustments, meaning to possibly develop a new set of sample models with higher accuracy and precision.

We have referenced some of these domains already in Figure 4-2 *AI information architecture overview diagram* in Chapter 4, "*AI Information Architecture*." In Figure 4-2, deployment and operationalization was categorized and depicted as a set of core ABBs. Furthermore, the ML workflow in the context of the core ABBs of the AI information architecture relates to some of these domains. One of the biggest challenges to operationalize AI is related to the required data engineering and pipelining tasks. This is caused by an often existing chasm between the AI development (including training, validation, and testing) and AI operational environments, where the circumstances and constraints of these two environments can greatly differ.

During AI development, for instance, data from different source systems may get consolidated, transformed, and pipelined; various techniques of feature engineering may get applied to eventually define the most suitable features for a particular ML or DL model. These pipelining and feature engineering tasks, however, may result into complex pipelines and data transformation modules, consuming a significant amount of time to build, often several days. During the operationalization of these ML/DL models and data engineering artifacts, these identical features have to be provisioned as input to the online scoring services in real time. This requires completely different technical integration and transformation capabilities, which more often than not cause unforeseen and tremendous data engineering and pipeline challenges for almost all enterprises.

---

[5]Please see Chapter 4, "*AI Information Architecture*," where we have described ML model accuracy and precision in the context of the ML workflow.

Table 6-1 lists the challenges of AI operationalization and productization, associated with the corresponding domain(s).[6] Needless to say that this list should be further customized and refined.

***Table 6-1.*** *Challenges for the Operationalization of AI*

| # | Challenge | Domain |
|---|-----------|--------|
| 1 | Ensure ML/DL model accuracy and precision throughout the entire life cycle | Model monitoring, model adjustments |
| 2 | Provide real-time access and data provisioning of required source data records for data engineering tasks | Data engineering |
| 3 | Provide real-time (low-latency) scoring services provided to applications and transactions | Model scoring, insight inference |
| 4 | Understand bias, fairness, and so forth of deployed ML/DL models | Model monitoring |
| 5 | Retrain models with new data sets to improve model relevance, accuracy, and precision | Model adjustment |
| 6 | Eliminate or reduce bias, overfitting, unfairness, and so forth in AI models, for example, via hyperparameter optimization or by applying new AI algorithms | Model adjustment |
| 7 | Exploit different platforms for AI model development vs. deployment, for example, public cloud vs. on-premises | Infrastructure transparency |
| 8 | Leverage different vendor offerings and open source tools for AI model development vs. deployment and operationalization | Infrastructure transparency |
| 9 | Chose different ML/DL algorithms to accommodate for changed use case characteristics or new available data | Model adjustments |
| 10 | Integrate and leverage open source tools and capabilities for ML/DL model operationalization (and development) | Infrastructure transparency |

(*continued*)

---

[6]This sequence of challenges is not within a particular order.

***Table 6-1.*** (*continued*)

| # | Challenge | Domain |
|---|-----------|--------|
| 11 | Apply required pipelining and data engineering tasks as preparation for scoring based on poor source data quality | Data engineering |
| 12 | Deliver required scalability and performance (e.g., low latency, real time) for all pipelining jobs | Data engineering |
| 13 | Explain and interpret the results, for instance, of scoring to adequately use insight within applications | Insight inference |

# General Aspects of AI Operationalization

In this section, we discuss the relationship and coherence of the following three general aspects of AI operationalization:

1. Deployment aspects

2. Platform interoperability

3. Vendor transparency

We are including AI model deployment aspects in the discussion insofar as this relates to the necessary and often cumbersome interchange of AI artifacts across various platforms, using various open source tools and libraries, as well as different vendor offerings. As we have mentioned at the beginning of this chapter, this requires advanced deployment capabilities, where the various tasks, such as AI model development including training, validation, testing, and deployment and versioning, may be performed with different tools and offerings installed and configured on different platforms.

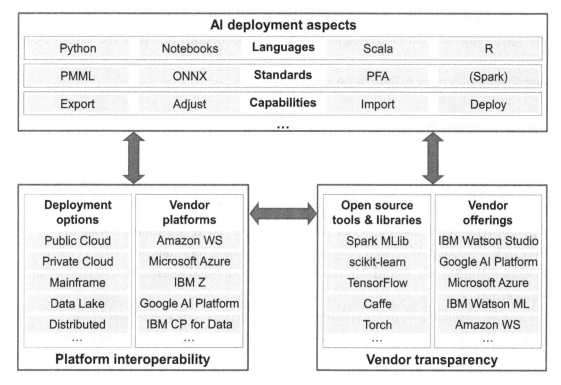

***Figure 6-1.*** *General Aspects of AI Operationalization*

Figure 6-1 is an illustration of the relationship or coherence of these three aspects. Prior to exploring these three general AI operationalization aspects individually, we would like to point out the influencing characteristics that these three aspects have on each other. The variations of AI deployment aspects, in terms of using different programming languages and standards and using the required deployment capabilities to, for instance, export and import AI models using different AI environments and frameworks, are driven by the reality of today's enterprises, which requires platform interoperability and vendor transparency.

In numerous engagements with IBM's customers, we have noticed the usage of many different open source tools, libraries, and frameworks and vendor offerings by the different enterprise organizations. This often leads to a heterogeneous shadow IT landscape, which prevents seamless vendor interchangeability and transparency to enable coexistence across organizations. Furthermore, different deployment options and vendor platforms stipulate the need for vendor platform and deployment interoperability. As you can see in Figure 6-1, the vendor transparency and platform interoperability depend on and influence each other.

In the following sections, we detail out deployment aspects, platform interoperability, and vendor transparency. It is recommended best practice to seriously take into consideration these three general aspects of AI operationalization and their interdependency prior to launching any AI endeavor.

# Deployment Aspects

As you have seen, AI deployments[7] can be characterized by using various programming languages, AI model-related standards, and a number of capabilities. We are limiting the set of AI deployment characteristics to those that are relevant for the interrelationship to platform interoperability and vendor transparency. There are quite a number of more or less popular AI-related programming languages, for example, Python (the most prominent one), Scala, R, and additional languages for notebooks, such as Python, Ruby, Perl, F#, and C#.

Several AI model-related standards[8] exist, where the Predictive Model Markup Language (PMML), Open Neural Network Exchange (ONNX), and Portable Format for Analytics (PFA) are some more prominent examples. Although Spark is not a standard, it is still popular among data scientists and provides a de facto standardization, especially for Spark ML model exchange and portability.

The applicability and choice of the many AI languages and standards depend on the use cases, skills, and preferences of individual organizations, data scientists, and data engineers. Some of the AI models can be transformed or integrated using a particular standard. Some AI models can also be transformed into other standards. Python scikit-learn models, for instance, can be put into production with PMML.

Deploying AI in the context of platform interoperability and vendor transparency necessitates several capabilities, such as exporting AI models from one specific vendor platform or vendor offering and importing them into another offering, which may be installed on a different platform, even with the need of some adjustments to be done to the AI models. For instance, IBM Watson Studio may have been used as part of IBM Cloud Pak for Data, which may be deployed on a private cloud platform to perform AI model development including training, validation, and test, with the deployment and operationalization for real-time scoring to be performed on an IBM Z mainframe environment.

---

[7]See [4] for more information on deploying AI, especially in regard to developing scorecards and performing comprehensive self-assessments.

[8]See [5] for more information on PMML, ONNX, and PFA.

This interoperability across platforms is vital for enterprises to accommodate the reality of use cases and organizational preferences and to optimize IT resource consumption.

## Platform Interoperability

Given the variety of different deployment options and vendor platforms available today, enterprises are often struggling with the right choice. Cloud deployments can be done in public, private, and hybrid or multicloud (mixed and multiple vendors) fashion. More often than not, today's enterprises entertain multicloud deployments across different organizations.

Data lake deployments – based on Apache Hadoop platforms – and distributed, for instance, Linux-based platforms enrich the number of possibilities. In recent years, traditional IBM Z mainframe systems, which still represent the IT backbone of the largest enterprises, have been modernized and significantly enriched with a set of ML and analytics-related capabilities[9].

As you have seen in Chapter 4, "*AI Information Architecture*," there are quite a number of vendor platforms, such as AWS, MS Azure, Google AI Platform, and IBM Cloud Pak for Data, which deliver – among other capabilities – AI-related services. These vendor platforms and deployment options need to interoperate with each other; they need to at least coexist and provide meaningful integration options, to enable the collaboration of roles and responsibilities, the interchange of AI artifacts, and meaningful split of AI-related tasks.

## Vendor Transparency

As we have mentioned at the beginning, there is a rich set of vendor offerings and open source tools, libraries, and frameworks to choose from. Vendor offerings[10], such as IBM Watson Studio, the Google AI Platform, Microsoft Azure, IBM Watson ML, and Amazon WS, are just a few examples.

---

[9]See [6] for more information on data and AI on IBM Z.

[10]See Chapter 4, "*AI Information Architecture*," where we have described some of these vendor offerings.

The list of AI open source tools, libraries, and frameworks[11] is sheer endless and confusing as well. Apache SparkML, scikit-learn, TensorFlow, Caffe, Torch, OpenNN, Theano, and Keras are just a few more prominent ones. Most vendors provide a rich choice and flexibility by integrating a number of open source packages into their offerings and thereby enabling organizations to choose the right libraries and frameworks in the context of their particular use case requirements. However, vendor transparency in terms of integrating different vendor offerings to, for instance, split AI workflow tasks across different tools and offerings and to deploy and operationalize AI artifacts across different vendor offerings still represents a challenge.

Prior to launching any AI endeavor, organizations should carefully evaluate vendor offerings in regard to their integrated open source tools, libraries, and frameworks and the integration capabilities across different vendor offerings. In order to avoid to be bound or locked into a particular vendor offering, transparency and openness should be examined. Thus, vendor transparency and openness is accompanied with platform interoperability and AI deployment aspects, as depicted in Figure 6-1.

# Key AI Operationalization Domains

As we have seen already, developing AI models, artifacts, and tasks, including training, validation, and test, involves quite a number of challenges. However, operationalizing AI and integrating AI artifacts and solutions into an existing application and IT landscape require enterprise readiness and a powerful and comprehensive AI information architecture. Enterprise readiness addresses the need for reliability, security, continuous availability, agility and flexibility, change management and life cycle management, and so on to operationalize AI solutions into a sustainable production environment. As we have discussed in Chapter 4, "*AI Information Architecture*," there are numerous ABBs that should collectively deliver a comprehensive infrastructure to develop and operationalize AI.

After we have discussed some general aspects, this section canvasses the following six key AI operationalization domains:

1. Data engineering and pipelining

2. Integrating scoring services

3. Inference of insight

---

[11]See [7] for a brief description of the most prominent AI open source tools, libraries, and frameworks.

4. Monitoring of AI models

5. Analyzing results and errors

6. Adapting AI models

Figure 6-2 (*key AI operationalization domains*) illustrates the relationship or coherence of these six domains or goals.

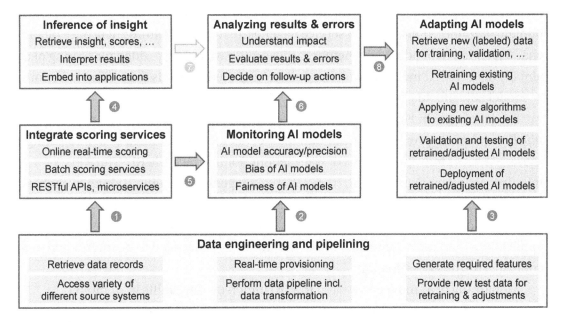

***Figure 6-2.*** *Key AI Operationalization Domains*

Prior to exploring these six key AI operationalization domains individually and in more detail, we would like to point out the influencing characteristics that these domains have on each other.

# Influencing Characteristics

In the context of operationalizing AI, the *data engineering and pipelining* domain plays a foundational role. This is caused by the key services that data engineering and pipelining needs to furnish to the following three domains: *integrated scoring services*, *monitoring AI models*, and *adapting AI models*[12].

---

[12]This is related to numbers 1, 2, and 3 in Figure 6-2, *key AI operationalization domains*.

Some of the data engineering and pipelining services, such as accessing a variety of different source systems to retrieve required data records, are very similar across the three domains mentioned earlier. Performing data pipelining, including data transformation tasks, is primarily related to the two domains *integrated scoring services* and *monitoring AI models*. Real-time feature provisioning is primarily geared to enable integrated scoring services, especially for online transactional scoring. However, it may also be leveraged to monitor precision and accuracy or bias and fairness of AI models in real time. The provision of new test data for retraining and adjustments is needed for the domain *adapting AI models*. The service to generate required features may have different characteristics, depending on whether the features have to be provided for scoring of an already deployed AI model or whether feature engineering needs to be done to perform AI model adjustments, by applying new ML algorithms.

As you can see in Figure 6-2 and the preceding discussion, the *data engineering and pipelining* domain provides key foundational services to operationalize AI. Integrating scoring services into applications or transactions, for instance, via RESTful APIs, microservices, or integration into transaction management systems, is done to enable inference of insight[13].

The outcome from the scoring services needs to be retrieved, embedded into applications, and turned into actionable insight. The results need to be interpreted in the context of the business domain.

Monitoring of AI models is geared toward understanding shifts in AI model accuracy and precision that may occur over time and to understand bias and fairness of AI models. To monitor AI models relies on scoring services to provide input, but also some data engineering and pipelining services[14] are required. The monitoring services serve as input to the *analyzing results and errors domain*, where results and errors need to be evaluated, and the impact (e.g., effort and cost, anticipated improvements, and business relevance) of possible adaptations needs to be understood, in order to decide on business-relevant follow-up actions. This may even require some input from the *inference of insight* domain[15].

---

[13]This is related to number 4 in Figure 6-2, *key AI operationalization domains*.

[14]This is related to numbers 2 and 5 in Figure 6-2, *key AI operationalization domains*.

[15]This is related to numbers 6 and 7 in Figure 6-2, *key AI operationalization domains*.

Once results and errors have been analyzed, *adapting AI models* is the AI operationalization domain, where AI models may be retrained with new labeled data, or they may be adjusted by, for instance, applying new ML models[16]. In some cases, a new set of AI models may even be developed, which means that a significant part of the AI workflow needs to be executed again. In any case, adapting AI models requires services from the *data engineering and pipelining* domain, for instance, to provide new labeled data for retraining and adjustments of AI models. If new ML algorithms may have to be used for AI model adjustments, then data pipelining, including data transformation, feature engineering, and so on, is required as well.

These six AI operationalization domains can be seen as a set of three AI operationalization workflows that are intertwined with each other. The first workflow is simply driving for inference of insight, the second one is to monitor AI models to understand business impact, and the third workflow is to react on this business impact by adapting chosen AI models.

After we have elaborated on the relationship and coherence, we move ahead by further exploring these six key goals individually.

# Data Engineering and Pipelining

Data processing and feature engineering, which is required for AI model and solution development is characterized by a specific set of challenges: source data need to be understood in the context of the business problem to be solved, the relevance of the data needs to be validated, and the data needs to be explored and visualized. Furthermore, the data has to be processed and labeled for AI models that need to be trained (potentially dealing with huge amount of data), evaluated, and selected. However, the time-critical aspects, performance characteristics, and data volume are significantly different during AI development compared to AI operationalization. In order to explore this further, let us take a closer look at the AI data engineering and pipelining workflow and set of tasks as depicted in Figure 6-3.

The left side of Figure 6-3 illustrates the data access, data processing, and feature extraction tasks. These tasks are very similar in nature, regardless whether the output is geared toward training and adaptation vs. operationalization including scoring and monitoring. Data needs to be accessed and converted (e.g., into a Spark DataFrame, a JSON format, or a simple CSV file), columns need to be dropped, data needs to be

---

[16]This is related to numbers 3 and 8 in Figure 6-2, *key AI operationalization domains.*

merged from different source systems, and data transformation and normalization tasks need to be performed. However, the context is different and the tasks may even have to be performed on different platforms: data processing for training and retraining of AI models doesn't require real-time or ultralow latency characteristics, as it is the case, for instance, for online transactional scoring.

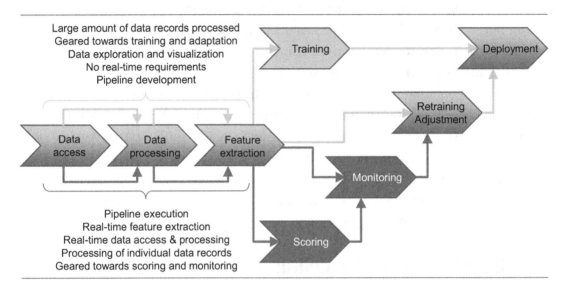

*Figure 6-3.*  *AI Data Engineering and Pipelining Workflow*

While developing AI models, all data processing steps can be conveniently performed using the available data engineering and data scientist platform, without time-critical aspects. When it is time for scoring (and also monitoring), the same data transformation and pipeline tasks need to be operationalized (productized) and performed in real time; this often causes huge problems. These time-sensitive aspects of provisioning required data from heterogeneous source systems for further processing and scoring in real time represent a challenge, which is often overseen or neglected during the development of the data engineering and pipeline modules. Even if some pipelines are an integral component of the so-called AI pipeline models, some data engineering tasks remain prior to feeding the data to the AI pipeline models for scoring.

Some of the data processing tasks, such as data exploration and data visualization, are unique to the AI model development phase; for AI operationalization, however, data doesn't have to be explored anymore, it needs to be processed in real time, and features need to be extracted in real time to provide the input for scoring of an AI model in ultralow latency fashion.

These specific characteristics of the entire data processing workflow are depicted in Figure 6-3 using different colors: the green color (upper part) for the AI development and the blue color (lower part) for the AI operationalization context.

It is recommended during the development of the AI data engineering and pipelining cycle to anticipate the operationalization aspects of these artifacts; AI operationalization aspects[17] must be taken into consideration right from the beginning of any AI endeavor.

## Integrated Scoring Services

Once AI models have been trained, validated, and deployed, they are ready to be exploited by the integrated scoring services, which is one of the key AI operationalization domains. As we mentioned at the beginning of this chapter, there are various ML and DL model types and interchange formats, such as Apache Spark models, scikit-learn models, and PMML or ONNX models, which also require different scoring engines. In addition to the different scoring engines, there are different ways in which the scoring service itself can be integrated into the IT infrastructure or requested by AI applications.

For instance, AI models can be scored in batch fashion or in real time; the integration can be done via RESTful APIs, within a transaction or an application, or integrated into a database management system (DBMS), for instance, by integrating the Spark or scikit-learn scoring engine into the DBMS. The particular option that a scoring service is integrated and requested by applications depends on the characteristic and specific requirements (e.g., performance) of a use case scenario.

As you have seen in Figure 6-3 (*AI data engineering and pipelining workflow*), the scoring of AI models requires the execution of data access, data processing, and feature extraction tasks. In other words, prior to the actual scoring of an AI model, at least some of the data engineering and pipelining tasks are assumed to have been executed already. As mentioned, some open source packages and vendor offerings allow the development of ML pipeline models, which can include a set of transformation steps prior to scoring a model. For instance, using the Apache Spark MLlib library, you can develop ML pipelines[18] that include transformer, which can either be an ML model or an algorithm that converts a Spark DataFrame data set into another one, for example, by adding a few

---

[17]See [8] for an example of AI operationalization of ML pipelines with PFA.
[18]See [9] for more information on Apache Spark ML pipelines.

columns. Depending on the requirements of a use case and the underlying source data systems, additional data engineering and pipelining tasks may have to be performed.

AI models can essentially be scored either in batch or real-time fashion. Batch scoring requires a set of source data records to be processed with the corresponding features to be extracted and supplied for the scoring of the AI model in bulk (batch) fashion. In this way, the required data engineering and pipelining tasks are performed in bulk fashion as well. Batch scoring can generate millions of scores to be interpreted by applications to, for instance, determine a list of clients to be targeted for a marketing campaign to be launched next week.

The following list describes some options regarding the implementation of *real-time* integrated scoring services:

1. **Online scoring engine integration**: The scoring services are integrated on a dedicated (stand-alone) application server. Applications can, for instance, request scoring services online via RESTful API scoring calls. Alternatively, even the same application server may be used for both the applications and the scoring services. Scoring services integrated on a dedicated (stand-alone) application server yield best performance. Implementations depend mainly on performance requirements and resources available.

2. **Transaction manager integration**: If there are performance requirements for high-throughput and large numbers of transactions per second in combination with an existing transaction management system, the scoring services should be integrated into the transaction management system. Applications that are running under the control of the transaction manager may then request scoring services via an interface module.[19]

3. **DBMS integration**: For use cases, where scoring of AI models can be linked to updates of data records initiated by events or applications, the scoring engine and service may even be integrated into the DBMS system. This is especially meaningful, if data transformation (e.g., aggregations, filtering, joins, etc.) is

---

[19]See [10] for an example of an integrated online scoring services.

required to generate the features for AI model scoring. Database triggers may be used to request scoring services.

4. **Application integration**: Specific privacy requirements and data sensitivity aspects may even suggest the integration of the scoring engine and service directly into the application.

Figure 6-4 depicts these four real-time integrated scoring scenarios on a conceptual level.

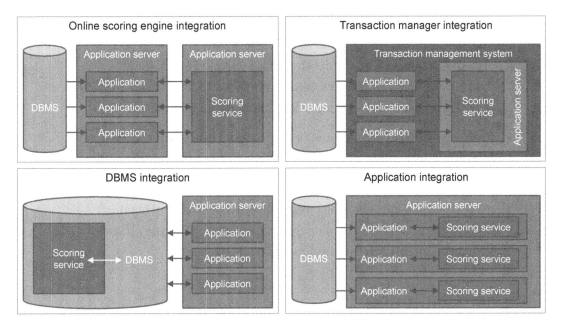

***Figure 6-4.*** *Real-Time Integrated Scoring Services*

There are a number of additional options and variations to integrate scoring services in real time, such as event processing and publication-/subscription-based implementations, which we won't discuss in this book.

# Inference of Insight

By inference of insight, we refer to the process of transforming the result of an AI model execution (e.g., a score, clusters, etc.) into actionable insight. This requires to retrieve and embed the outcome of the AI model execution into an application, to interpret the results (e.g., a score) within a given business or use case context, and to decide often in real time (or with ultralow latency) on follow-up actions. Thus, inference of insight aims to understand the implications and conclusions that can be derived from the outcome of an AI model execution, which is strongly related to the business problem, and has to be integrated to the greatest possible extent into the business application.

The following are some key capabilities to enable inference of insight[20]. Depending on the business context or use case scenario, these capabilities may have to be applied in real time.

- **Interpretability**: The ML/DL model outcome has to be easily verifiable and interpretable, meaning comprehensible, given a certain business or use case context.

- **Trustworthiness**: The results should be reliable and trusted by consumers and applications without the need to further scrutinize or challenge the results with regard to their applicability.

- **Consumability**: The outcome of an AI model execution needs to be consumable, meaning easily embeddable into consuming business applications.

- **Accountability**: Depending on the business context and legal implications of an AI-based decision, the applications may have to take accountability requirements into consideration.

# AI Model Monitoring

AI models are developed within a particular business context. Business and use case characteristics or measures, however, may shift over time. The data itself may be drifting over time as well. These changing circumstances may result in reduced business relevance of the AI application and a decline in accuracy and precision of the AI models. In addition, AI models may not yield the results that were originally anticipated; they may show certain anomalies only later in their life cycle, be less fair, and show strong undesirable bias. In order to understand this decline and to take appropriate action, corresponding AI model measures need to be constantly calculated and monitored.

---

[20]We would like to point out again that the term *inference* is often used for a subset of the AI operationalization scope as used in this book.

In Chapter 4, *"AI Information Architecture,"* we have already highlighted model monitoring as an essential component, especially as part of the ML workflow, and as one of the ABBs related to deployment and operationalization. In this section, we detail out the various areas that should be monitored – and why. The scope, however, is limited to the monitoring of the AI model outcome, not the broader scope of, for instance, traceability and auditability of AI assets related to the entire life cycle.[21]

The following list describes some of the key measures of AI model monitoring. Business requirements and the specific needs of a particular use case scenario determine the applicability of either all or a subset of these measures. In addition, for some use case scenarios, offline model monitoring may be sufficient; however, for time-sensitive use cases, where circumstances and characteristics may drift frequently, online AI model monitoring may be required.

1. **AI model performance**: Monitoring of ML and DL model performance mainly relates to measuring the area under the receiver operating characteristic (ROC) and precision-recall (PR) curves for AI regression and classification models. Declining areas under the ROC curve (measuring the accuracy of an AI model) and PR curve (measuring the precision of an AI model) can indicate a worsening of the AI model accuracy and precision, which – depending on the use case – means worsening business measures, such as losing money, declining client satisfaction, and so on. Even the shape – meaning the balance of the corresponding measures – of the ROC and PR curves may have a business impact. For credit card fraud models, for instance, where the PR curve shows higher recall numbers (true positive rate) and lower precision numbers (ration of true positives to predicted positives), this is an indication for less false negatives, meaning less fraudulent cases not discovered. If the PR curve has a swing toward higher precision and lower recall numbers, this is an indication for less false positives, meaning, for instance, that less credit cards are retained – although the area under the PR curve is identical in both cases. Monitoring of AI model performance requires additional measures that depend on the requirements

---

[21]Please refer to Chapter 8, *"AI and Governance."*

and characteristics of a particular use case scenario. For instance, AI model scoring or outcome that serves as a base for decisions and actions as part of an AI application (e.g., medical diagnosis and treatment, self-driving vehicles) may have to be monitored holistically. That is to say, for AI model performance, monitoring needs to be augmented to the decision optimization level, which may even take human intervention into consideration. For self-driving vehicles, for instance, AI modeling and DO need to be autonomous[22] and adjustable to personal driving behavior and individual driver preferences – and even the cultural context.

2. **Bias and fairness of AI models**: An AI model may not yield the desired result in terms of showing too much bias and lack in fairness. Monitoring of AI model results is required to understand biased outcome, meaning an AI model to show favorable or unfavorable outcome related to a specific feature or a subset of features. Furthermore, the monitoring may lead to further analysis and visualization of the impact of an adjusted AI model with adjusted feature weights to mitigate bias. For regression or classification models, a certain degree of bias may be a desirable outcome. For instance, if the business problem with its applied labeled data set for training, validation, and testing is simply suggesting a favor for a certain gender (e.g., female) or age cluster (e.g., $30 \leq age \leq 39$), or color (e.g., blue), then a bias toward these features is completely normal and even desired. However, circumstances and new data may change over time, which may result in unfavorable bias. This shift in bias needs to be detected and acted upon. Vendors are already providing offerings to detect bias, including root cause analysis.[23]

3. **Explainable outcome of AI models**: The outcome of an AI model and decisions of AI applications need to be explainable and even visualized, meaning to explain and visualize model decisions to a business user in business terms. Key business

---

[22]See Chapter 13, *"Limitations of AI,"* where we elaborate more on autonomous ML and DL.
[23]See Chapter 13, *"Limitations of AI,"* for more information on bias.

measures or KPIs (e.g., number of fraudulent cases, amount of money lost, number of credit cards retained, number of customers complaining because of erroneously retained credit cards) need to be correlated to the metrics of an AI model (e.g., accuracy or precision metrics, shape of the ROC and PR curves), so that monitoring of AI model performance figures can be put into a meaningful business context and acted upon. Specifically for AI with autonomous, meaning continuous automation of learning, where self-governing processes with limited or even no human intervention are prevalent, explainability of AI models and applications is reaching a new dimension.[24]

In today's enterprises, AI model monitoring is still done in a rather rudimentary way, with limited support through open source tools and vendor offerings to enable automation. Although measurement and monitoring of technical AI model performance measures – for instance, the area under the ROC and PR curves – can be implemented, the correlation of AI model results to business-relevant KPIs is a rather emerging area with much still left to be done.[25]

# Analyzing Results and Errors

For those cases, where AI model performance deterioration, bias and unfairness, or additional undesirable occurrences have been detected, they have to be analyzed and the implications have to be understood. Some of the results and errors may not have such a significant impact on the business, particularly if measured against the anticipated effort of possible follow-up activities. As you can imagine, these activities need to be tailored according to the specifics of your use case scenario. In some cases, the AI application itself with its decisions based on certain measures (e.g., scores with probability figures) may have to be adjusted – not necessarily the AI model.

---

[24]See Chapter 13, "*Limitations of AI*," where we elaborate more on the explainability of decisions.
[25]See [11] for an example of measuring outcome from AI against business KPIs.

The following is a short list of the key tasks of this AI operationalization domain:

- **Evaluate results and errors**: The monitoring results, occurrences, and phenomena need to be evaluated.

- **Understand the impact**: Their impact to the business needs to be understood prior to determining any possible follow-up actions.

- **Decide on follow-up actions**: Depending on the effort, certain follow-up activities, such as model retraining, may be advisable.

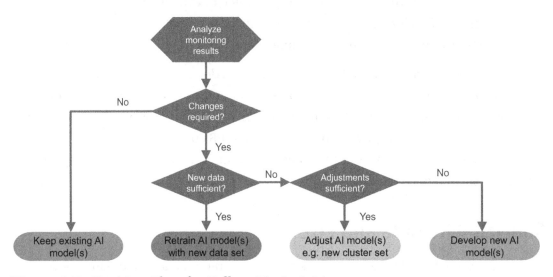

***Figure 6-5.*** *Decision Flow for Follow-Up Activities*

Figure 6-5 is a simple decision flow regarding possible follow-up activities. As you can see in Figure 6-5, there may very well be situations where the existing AI models should not be changed. The AI application(s) may have to be adjusted instead, for instance, by adjusting levels of probability figures for scores that the application needs to act upon. In other cases, retraining of the existing models using new labeled data sets is sufficient. In those situations where the drift of data records is not causally responsible for the worsening of the AI models, adjustments of the models or even the development of new models may be advisable.

We now briefly discuss the three possible AI model adaptation options in the next section.

# AI Model Adaptations

Once the decision has been made to adapt an existing AI model, there are essentially three meaningful options, as we have seen in the previous section: (1) retraining an existing AI model with new labeled training, validation, and test data; (2) adjusting an AI model, which means to possibly further optimize hyperparameters or, for instance, in the case of classification models, to adjust the numbers of clusters; and (3) developing a new AI model by evaluating a new set of ML/DL algorithms.

1. **Retrain existing AI model(s)**: More straightforward situations exist, if there are moderate changes to the underlying business model or data records, which may simply be addressed by retraining the existing AI model(s) by using new labeled data sets. In such cases, the effort involved should be reasonable. This effort comprises the provisioning of new labeled data sets for training, validation, and testing; and the deployment of the retrained AI models.

2. **Adjust existing AI model(s)**: In those cases, where the monitoring results show a more significant impact to the business, which is not addressable by simply retraining the existing AI models, adjustments to the number of clusters or the number of levels and nodes of a decision tree model (to increase the information gain) and further optimization of hyperparameters have to be performed. These adjustments may already imply more significant development effort for data scientist and engineers.

3. **Develop new AI model(s)**: If the changes of the business situation and the use case circumstances are so serious that hyperparameter optimization or adjustments to the existing AI applications and AI model(s) as described previously may not be promising, developing a new set of AI model(s) may be advisable. This, of course, equates to a more comprehensive development effort that comprises all steps of the AI workflow, involving all personas and stakeholders, such as business users, data scientists, and engineers.

# Key Takeaways

We conclude this chapter with a few key takeaways, as summarized in Table 6-2.

***Table 6-2.*** *Key Takeaways*

| # | Key Takeaway | High-Level Description |
|---|---|---|
| 1 | Operationalizing AI represents a huge challenge | It is especially related to the required efficiency of AI model deployment, real-time model pipelining and scoring, auditing, and monitoring |
| 2 | Need for platform interoperability and vendor transparency | Platform interoperability (including deployment options and vendor platforms) and vendor transparency (including open source and vendor offerings) are required for flexible AI productization and operationalization |
| 3 | AI operationalization domains influence each other | These are six domains: data engineering and pipelining, integrated scoring services, inference of insight, monitoring AI models, analyzing results and errors, and adapting AI models |
| 4 | Differences in AI data engineering workflow | There are key differences of the AI data engineering and pipelining workflow during development vs. operationalization |
| 5 | There are a number of integrated scoring services options | There are four key scoring services options: online scoring engine, transaction manager, DBMS, and application integration |
| 6 | There are required capabilities for inference of insight | We have identified four capabilities: interpretability, trustworthiness, consumability, and accountability |
| 7 | AI models need to be monitored | To ensure AI model performance (accuracy and precision), to detect bias and ensure fairness, and to generate explainable outcome |
| 8 | AI model monitoring results need to be analyzed | For all occurrences, the impact to the business needs to be understood, and possible follow-up activities have to be performed |
| 9 | AI models may have to be adapted | Adaptations could mean to retrain existing AI models with new labeled data sets, to adjust existing AI models, for instance, via hyperparameter optimization, or to develop a new set of AI models |

# References

[1]    Jyoti, R. *Accelerate and Operationalize AI Deployments Using AI-Optimized Infrastructure*. IDC. IDC Technology Spotlight, 2018, `www.ibm.com/downloads/cas/ZYGVAOAL` (accessed March1, 2020).

[2]    Walch, K. Forbes. *Operationalizing AI*, 2020, `www.forbes.com/sites/cognitiveworld/2020/01/26/operationalizing-ai/#49ef691c33df` (accessed March 2, 2020).

[3]    IBM. IBM Knowledge Center, *Configuring WML for z/OS scoring services in a CICS region*, 2020, `www.ibm.com/support/knowledgecenter/SS9PF4_2.1.0.2/src/tpc/mlz_configurescoringservicecics.html` (accessed March 2, 2020).

[4]    Blokdyk, G. *Deploying Artificial Intelligence – A Complete Guide*. ISBN-13: 978-0655810681, 5starcooks, 2019.

[5]    Levitan, S., Claude, L. *Open standards for deployment, storage and sharing of predictive models. PMML / PFA / ONNX in action*. Conference: Applied Machine Learning days 2019, DOI: 10.13140/RG.2.2.31518.89920, 2019, `www.researchgate.net/publication/334611859_Open_standards_for_deployment_storage_and_sharing_of_predictive_models_PMML_PFA_ONNX_in_action` (accessed March 6, 2020).

[6]    IBM. *Data and AI on IBM Z — insight at the point of interaction*, `www.ibm.com/in-en/analytics/data-and-ai-on-ibm-z` (accessed March 6, 2020).

[7]    Deb, A. Edureka. *Top 12 Artificial Intelligence Tools & Frameworks you need to know*, 2019, `www.edureka.co/blog/top-12-artificial-intelligence-tools/` (accessed March 7, 2020).

[8]    Pentreath, N. IBM. *Productionizing Machine Learning Pipelines with PFA*, `https://events19.linuxfoundation.org/wp-content/uploads/2017/12/Productionizing-ML-Pipelines-with-the-Portable-Format-for-Analytics-Nick-Pentreath-IBM.pdf` (accessed March 10, 2020).

[9]    Apache Spark. *MLlib Main Guide. ML Pipelines*, `https://spark.`
`apache.org/docs/latest/ml-pipeline.html#dataframe`
(accessed March 11, 2020).

[10]   IBM. IBM Knowledge Center, *Configuring WML for z/*
*OS scoring services in a CICS region*, `www.ibm.com/`
`support/knowledgecenter/SS9PF4_2.1.0/src/tpc/mlz_`
`configservicesincics.html` (accessed March 12, 2020).

[11]   IBM. *IBM Watson OpenScale.* `www.ibm.com/cloud/watson-`
`openscale/` (accessed March 14, 2020).

# CHAPTER 7

# Design Thinking and DevOps in the AI Context

Design thinking and DevOps are long-standing concepts, well established and leveraged by most leading organizations. A sustainable adoption of AI for these concepts, however, is still lacking. But what exactly do we mean by adopting AI for design thinking and DevOps, and what possibilities do we have anyhow? Design thinking and DevOps methods can certainly be applied to develop AI systems and devices, products and tools, or applications. This is probably a more obvious thought. But can AI and its siblings be leveraged and infused into design thinking and DevOps concepts – and how? What are the prerequisites, challenges, and the benefits to do so? An obvious prerequisite is to first establish a sound design thinking and DevOps infrastructure and culture, prior to introducing AI and ML into these concepts.

This chapter attempts to answer the preceding questions. We begin by revisiting the key concepts of design thinking and DevOps and provide some ideas for design thinking and DevOps in the context of AI. We furthermore describe key aspects of AI design thinking and AI DevOps, focusing on the enrichment of design thinking and DevOps by infusing AI technologies and data science methods.

## Introduction

As mentioned, design thinking and DevOps have found their way into most leading corporate cultures. This is especially true for design thinking methods; however, DevOps seems to be artificially applied to all applications, middleware products, tools, and solutions, regardless whether it is essential and meaningful or not. For instance, DevOps is essential for systems of engagements, where user experience needs to be continuously adjusted and improved, but less meaningful for systems of records or even systems of

© Eberhard Hechler, Martin Oberhofer, Thomas Schaeck 2020
E. Hechler et al., *Deploying AI in the Enterprise*, https://doi.org/10.1007/978-1-4842-6206-1_7

insight, where new product or tool features and functions that are invoked much too often may result into a cumbersome integration endeavor causing a rather disrupting experience for operational staff.

When it comes to AI technologies and data science methodologies, we need to understand the bidirectional nature of influence by applying design thinking and DevOps methods to AI and also applying AI to these methods.

On one side, AI technologies with ML and data science methodologies can and should be leveraged by design thinking and DevOps methods in order to enrich and improve these concepts. For instance, AI and ML can be used to introduce predictive qualities for the design (e.g., predicting the level of acceptance by new or existing user clusters) or to discover patterns in the user data, leading to qualitative and quantitative design evaluations. AI for DevOps can contribute to self-governed systems or to resource optimization in DevOps.

On the other side, design methods and DevOps – assuming that these concepts are well established – can be applied to the design, development, test, deployment, and operations of AI solutions and devices. For instance, AI methods can be used within the design thinking process to provide designers with more relevant and user-centric insight or to discover correlations during the design thinking ideation phase. For DevOps, AI can discover patterns and root cause of system outages, leading to accelerated and more targeted feedback from operations to development teams.

# Design Thinking and DevOps Revisited

This section revisits traditional aspects of design thinking and DevOps and provides a definition as well as a high-level overview of design thinking and DevOps concepts, processes, approaches, and the benefits.

## Traditional Design Thinking

Design thinking is not a brand-new concept; it's around for decades. Originally, it was used for new product development, not even limited to science, engineering, or programming. The interdisciplinary character and user-centricity of design in general was already an emerging theme in architecture and industrial design in the 1920s. With its long history and broad application in business and industry, architecture and engineering, and science (including computer science), there is no single commonly accepted definition of design thinking. For the purpose of our discussion, design thinking[1] is a human-centered approach to design that is comprised of processes and methods to foster creative and innovative out-of-the-box thinking to solve complex problems.

The core concept of design thinking with the idea for rapid prototyping and fast, interactive user feedback is based on the following principles:

- **User-centricity**: Focus during the design and system development cycle is on the need of the end user or customer, ensuring increased usability and driving for high user acceptance and pleasant user experience – enabling user-convergent thinking.

- **Multidisciplinary**: Collaboration is across various disciplines and personas (e.g., entrepreneurs, design, engineering and programming, marketing, psychology, lawyers) embracing different perspectives – ensuring a holistic and pervasive approach.

- **Democratizing participation**: Environment (e.g., rooms, facilities, etc.) and methods empower and encourage teams with *all participants* to contribute with equal appreciation and nonjudgmental attitude – fostering creative and rich ideation.

There are a number of additional principles of design thinking, such as defining smaller-scale, meaningful, and achievable targets for an intermediate user outcome, collaborating with sponsor users to receive and incorporate feedback, and establishing an agile and lean design thinking process that continuously improves solutions in the context of new emerging problems.

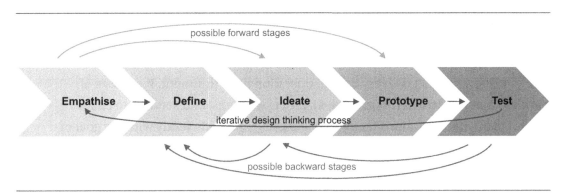

*Figure 7-1.* *Design Thinking Process*

---

[1]See [1] and [2] for more information on design thinking.

Several design thinking processes with different stages have emerged in the past. One of the original ones was proposed already in 1969.[2] It consists of the following seven stages (steps or phases): define ➤ research ➤ ideate ➤ prototype ➤ choose ➤ implement ➤ learn. Another process is comprised of only three stages with corresponding substages:[3] inspiration (understand, observe, PoV) ➤ ideation (ideate, prototype, test) ➤ implementation (storytelling, pilot, business model). We are using the design thinking process with five stages as depicted in Figure 7-1. The five stages are as follows:

1. **Empathize**: Gain deep understanding regarding the problem to be solved, and empathize with end users or customers.

2. **Define**: Structure information that was gathered around the problem, and apply your insight and understanding.

3. **Ideate**: Develop innovative ideas and design points regarding the solution and its capabilities.

4. **Prototype**: Create first solution as a prototype with limited functions and features as input for test.

5. **Test**: Perform test and validation of the prototype, and incorporate feedback from sponsor user testing.

It is important to notice that the stages don't necessarily have to be performed strictly in sequence: stages can be skipped (depicted in Figure 7-1 as forward stages) or performed in parallel. Backward stages are possible as well, especially to incorporate test results or sponsor user feedback into the define or ideate stages.

# Traditional DevOps

DevOps[4] is geared to assimilate and blend agile development and operations to accelerate new product delivery into an operational environment. It consists of methodologies, practices (or principles), processes, tools, and services to enable development and operations teams to collaborate more efficiently with the goal of seamless and more frequent deployments.

---

[2]See [3] for more information on the original design thinking processes from Herbert A. Simon.
[3]See [4] for more information on this design thinking process.
[4]See [5] and [6] for more information on DevOps.

As we mentioned in our introduction to this chapter, DevOps should not be misunderstood as a frequent disruption to IT operations. For instance, migrating to a newer product version can be a complex and time-consuming process, involving countless assumptions and prerequisites to be met. Systems of record and insight are often integrated within a rather complex IT infrastructure, with significant gaps regarding DevOps processes, tools, and services to guarantee an Ops-friendly experience – even today. We therefore see the applicability of DevOps more related to systems of engagements.

We come back to this particular aspect when we elaborate on DevOps in the context of AI and AI for IT operations (AIOps).

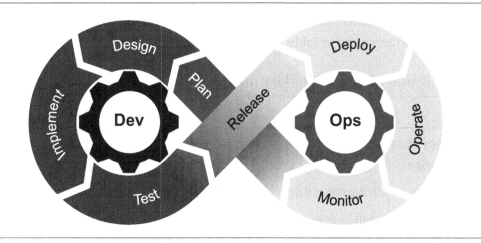

***Figure 7-2.***  *DevOps Life Cycle*

Figure 7-2 illustrates the process interleaving of development with operations, leading to the DevOps life cycle, which needs to be performed by people and systems and should be underpinned with methodologies and principles, processes, and tools that provide seamless integration and required automation for all stages as depicted in Figure 7-2.

The core concept of DevOps is based on the following principles:

- **Continuous integration**: Enable lean, agile, and reliable software development, delivery, deployment, and operations processes and tools that are aligned with each other.

- **Intensified collaboration**: Foster development and operations teams to collaborate in order to break out of siloed organizational thinking, moving toward a holistic system approach.

- **Ongoing improvement**: Encourage ongoing learning and continuous feedback across the organization to implement improvements based on best practices and tools.

There are quite a number of DevOps success factors and reasons of failures, which should also be factored in. Gartner[5] *"predicts that through 2022, 75% of DevOps initiatives will fail to meet expectations due to issues around organizational learning and change."*

# Benefit of Design Thinking and DevOps

Before shifting the focus toward AI again, we briefly list the benefits of traditional design thinking and DevOps. Since some of the benefits are driven by both concepts[6], we summarize this within a single Table 7-1:

*Table 7-1.* *Benefit of Design Thinking and DevOps*

| # | Benefit | Concept |
|---|---------|---------|
| 1 | Encourages interdisciplinary approach, accelerating creativity and innovation | Design thinking |
| 2 | Focus on problem solving, rather than technical feasibility | Design thinking |
| 3 | Faster code and problem fix delivery | DevOps |
| 4 | Increased SW and application quality | DevOps |
| 5 | Improved collaboration across often siloed organizations with faster time to market | DevOps and design thinking |
| 6 | Enhanced participation of *all* team members | Design thinking |
| 7 | Improved user-centricity and user experience | Design thinking |
| 8 | Optimized and more efficient responsiveness to market and user demands | DevOps and design thinking |
| 9 | Iterative approach enables continuous learning, knowledge, and product and operational improvements | DevOps and design thinking |
| 10 | Reduced complexity through smaller components | DevOps |

---

[5]See [7] for more information on Gartner.
[6]See [8] for more information on the interaction and coexistence of design thinking and DevOps.

# Design Thinking in the Context of AI

Are we considering design thinking for AI or AI for design thinking? Design thinking methods should obviously and undeniably be applied to the design and development of AI solutions and devices.[7] This will naturally ensure AI to become much more human- and user-centric. This section, however, is devoted to the increased influence of AI technologies and data science methodologies on design thinking itself. That is to say, we are concerned about how does design thinking change under the influence of AI, and what are the possible improvement areas?

# AI Influence on Design Thinking

The scope of the following ideas is equally relevant for applying AI-infused design thinking methods to solving challenging problems, as well as designing and developing new products, tools, or services.

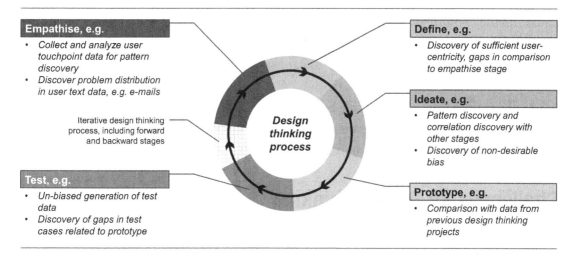

**Figure 7-3.** *AI-Infused Design Thinking*

There is a vast amount of end-user and customer touchpoint data available, for example, online browsing and cookies data, email threads and complaints records, App problem reports, historical purchasing data and customer profile information, location data, and so on. Especially the *empathize, define*, and *ideate* stages of the iterative design

[7]See [9] for more information on applying design thinking to AI.

thinking process, as depicted in Figure 7-3, will receive more precise and deeper input based on AI-derived insight from this data. For instance, this essential consumer and end-user touchpoint data can be used by data scientists to develop ML and DL models, understand customer decisions, predict customer behavior, and discover correlations of product or service usage to complaints records or App problem reports. This analytical insight becomes increasingly relevant to improve and live up to the aspiration of the high user-centricity of design thinking, eventually delivering an increased user experience.

Data and information that is relevant for the *ideation* and even the *prototype* stages of the design thinking process can be analyzed and used to predict qualities, relevance, dependencies, and other design measures (KPIs) of ideas and design points. Data scientist methods with ML and DL models provide designers and the entire interdisciplinary team with more relevant and targeted insights, therefore increasing confidence and accelerating the various design thinking process stages.

Design thinking can benefit from analyzing data from all process stages of past projects to discover anomalies, patterns, and correlation from data of different stages to predict project execution based on new data from current process stages. For instance, data from the *empathize, define,* and *ideate* stages can be analyzed with data science methods to discover nonobvious correlations, such as bias of certain ideas or design points specified during the *ideate* stage toward specific input data or structures from the *empathize* or *define* stages.

AI can also be leveraged to discover gaps or patterns of the *test* stage, as it relates to the prototype code. Even test data generation can be optimized via AI to ensure adequate coverage of defined scenarios.

# Challenges for Design Thinking

As we have seen in the previous chapters, there are numerous far-reaching challenges in adopting AI. Table 7-2 strictly focuses on those challenges that are relevant in adopting AI for design thinking.

***Table 7-2.*** *Challenges for Design Thinking*

| # | Challenge | Description |
|---|-----------|-------------|
| 1 | Design thinking approach in place | Infusing AI requires an existing, sound, and mature design thinking approach within the organization – this, however, may not always be the case |
| 2 | Complexity of AI | AI may be seen as too sophisticated and complex to be adopted within design thinking methods |
| 3 | Skill and knowledge gaps | Team members have insufficient AI and data science skills and knowledge |
| 4 | Lack of AI-infused tools | Design thinking tools may still lack sufficient AI capabilities |
| 5 | Adequate data for learning | ML models require adequate data for learning (training) and validation, which may be a challenge to get |
| 6 | Skepticism within teams | The multidisciplinary teams may not have accepted AI as a vehicle to improve design thinking |

# DevOps in the Context of AI

Similar to the opening question in the section "*Design Thinking in the Context of AI*," we need to ask here as well whether we are considering DevOps for AI or AI for DevOps. Of course, DevOps can and will be applied to the interleaved cycle of development with operations in regard to AI solutions and devices. However, in this section, we focus on the required adaptation and enhancements of the DevOps stack because of the application of AI. In other words, what are the required changes for DevOps, for instance, to factor in AI artifacts, such as ML and DL models, and what are meaningful and desirable DevOps enhancements to take advantage of AI and data science methods?

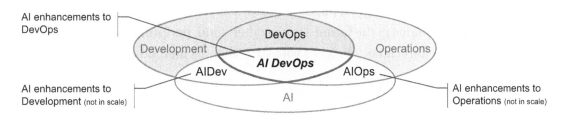

***Figure 7-4.*** *Relationship of AI DevOps with AIOps and AIDev*

Before we carve out the influence of AI on DevOps, we would like to clarify the terms AIOps and AIDev, as they are depicted in Figure 7-4. AIOps is the adaptation of AI for operational processes. Thus, you should view AIOps as an enhancement to traditional operations, for instance, IT operations (sometimes called ITOps). The same applies to AIDev, which is the adaptation of AI for traditional development processes. Thus, AIDev is the enhancement of development by applying AI methods and techniques. DevOps – as we have seen earlier – is the agile relationship between development and operations. AI for DevOps – which we call AI DevOps – can be seen as an enhancement to DevOps via the adaptation of AI methods and techniques to the traditional DevOps area. Therefore, AI DevOps intersects with AIOps and AIDev in those areas, where the infusion of AI for operations and development is relevant for DevOps purposes. Figure 7-4 is a high-level depiction of the relationships of AI DevOps with AIOps and AIDev.

# AI Influence on DevOps

DevOps and the broader area of IT operations (ITOps) are ideal domains for applying AI; however, we focus "only" on DevOps. AI imposes new challenges for DevOps. For instance, ML and DL models require new deployment and operationalization capabilities that have an impact on DevOps.[8] AI DevOps can naturally be leveraged to separately improve the development and operations side. For instance, on the development side, AI can be used for segmentation and improved automation of test cases. On the operations side, AI can be used for baseline determination and anomaly detection of one or several SW subsystems (e.g., database or transaction management systems).

However, the sweet spot for AI to augment DevOps is to target the intersection of development and operations with the goal to further optimize the DevOps cycle as depicted in Figure 7-2. For instance, log data from the operational environment can be used to develop ML models to discover correlations between application failures and subsystem configurations, which can be used by the development teams to optimize their application features or suggest improved subsystem configurations with the goal to limit application failures in the future. On the other hand, performance test data from development teams can be analyzed with data scientist methods to recommend

---

[8]Please recall the key messages from Chapter 6, "*The Operationalization of AI.*"

deployment options to support operational teams to optimize deployment and operational aspects of a new application version.

Continuous development, deployment, integration, and operations require AI to accelerate and automate the DevOps cycle. The following (incomplete) list contains some additional AI application opportunities to improve and accelerate the DevOps cycle:[9]

1. **Anomaly detection**: Anomalous application behavior can be detected by ML and fed back to development in a timely manner to provide automated fixes to shorten the window of impact.

2. **Performance issues**: Predicting performance bottlenecks via correlations of performance-relevant KPIs can alert development to optimize software for more efficient resource consumption.

3. **Suggest optimization**: Using AI, development can suggest optimized deployment and configurations and adequate parameter settings for different types of workloads.

4. **Test optimization**: Continuous testing and AI-based test case selection can reduce and streamline testing and accelerate development, delivery, and deployment.

5. **Module identification**: Historical data from development and operations can be used to determine relevant SW modules in the context of undesirable system or application behavior or errors.

6. **Crossing silos**: Correlation discovery across different IT infrastructure components (e.g., memory utilization) and number of users can help development to optimize products in regard to resource consumptions.

There are additional forward-thinking ideas, for instance, to use AI to translate functional specifications written in natural language into executable code segments or to use AI to automatically transform undesirable operational behavior into *functional feature requests* for development.

---

[9]See [10] and [11] for more information on using AI in the context of DevOps.

# Challenges for DevOps

Despite the convincing list of AI application opportunities for DevOps, you may be faced with some challenges, as outlined in Table 7-3. Considering the fact that development and operations teams are often separated by enterprise boundaries, these challenges become even more substantial.

***Table 7-3.*** *Challenges for DevOps*

| # | Challenge | Description |
|---|-----------|-------------|
| 1 | Sound DevOps approach in place | Infusing AI requires a sound existing DevOps approach already implemented and accepted in your organization(s) and the corresponding client organizations |
| 2 | AI-infused tools not mature enough | AI DevOps requires tools available on the development and operations side; they need to be integrated and geared toward the DevOps cycle |
| 3 | AI design thinking may slow down the process | There may be a perception that AI design thinking may significantly slow down the process – at least initially. AI design thinking may indeed require an additional investment in tools, learning, and adjustment of the existing design thinking process |
| 4 | Adequate data for learning | Similar to design thinking, ML models for AI DevOps require adequate data for learning (training) and validation, which may be a challenge to get |
| 5 | Skill and knowledge gaps | Team members have insufficient AI and data science skills and knowledge |
| 6 | AI complexity too high | Skepticism may be significant simply because of the complexity of AI, and hesitation of teams to adopt AI has to be addressed |
| 7 | Required data not available | For AI to become essential for DevOps, quite a lot of data from dispersed, often siloed systems has to be collected, transformed, and explored, which requires corresponding tools and willingness to share data across organizations |

# Key Aspects of AI Design Thinking

This section is devoted to the key aspects and the value of AI design thinking. Our intention is to develop a simple AI design thinking framework or model that maps the most noticeable AI features to the design thinking process as depicted in Figure 7-1. A short discussion on the value of AI design thinking concludes this section.

# AI Design Thinking Model

As we have stated in Table 7-3, *Challenges for DevOps*, there may be the perception that AI in the design thinking process is difficult because of the time involved to do so. Implementing an AI design thinking model may indeed require an investment to further improve the characteristics of design thinking regarding rapid development, prototyping, testing, and improvements. Our AI design thinking model[10] is limited to an AI feature mapping exercise, deliberately leaving out other aspects of a model, such as an architecture overview diagram with component interaction diagrams, a comprehensive data flow description, and so forth. Nevertheless, this narrow view of depicting key AI capabilities in the context of the design thinking process helps you to gain a more structured understanding of AI design thinking and enables you to appreciate the value that can be derived from AI.

To begin with, we need to agree on key AI features that are meaningful for our exercise. To limit these features to a manageable size, a certain level of abstraction is of the essence. We are limiting our discussion to the following AI capabilities.

- **Data preparation**: With data access, exploration, and visualization

- **ML**: With predictive analytics, pattern, and correlation discovery

- **Text and voice**: With text analytics, sentiment analysis, and NLP

- **DL**: With ANNs, image recognition, and video processing

There are without doubt numerous additional AI capabilities, such as robotics, and cognitive areas other than NLP (e.g., reasoning), rules management, planning, and scheduling, which we are deliberately leaving out. Image recognition and video processing are listed under DL for the sake of completeness.

---

[10]See [12] for more information on design related models and frameworks.

As you can see in Figure 7-5, we have depicted the key AI capabilities in four groups and have mapped those capabilities exemplarily to the *ideate* stage of the design thinking process. We briefly provide additional ideas regarding the mapping of AI to the remaining four stages.

The *empathize* stage will particularly focus on the voluminous end-user or customer touchpoint data. To ensure user-centricity is without any doubt the vital objective for the empathize stage. Thus, data exploration and visualization are essential to understand the relevance of data records and to limit the data volume for subsequent processing steps. Data science techniques can be used to discover patterns in the data, to perform clustering, to understand priorities, and to detect correlations that can serve as input to subsequent stages.

Since the majority of the data may be unstructured, text analytics can be brought in as well.

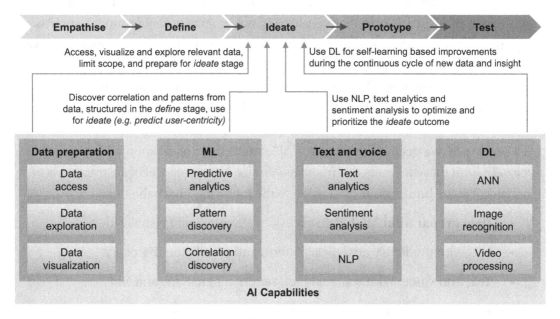

***Figure 7-5.*** *AI Design Thinking Model*

Insight from the *empathize* stage enables the structuring of data as part of the *define* stage. Predictive analytics can be used to gain further insight regarding relevant measures and projections of possible design points. In other words, predictive analytics with ML models increases confidence to drive for required user-centricity and enables prioritization of input to the *ideate* stage.

The *prototype* stage can benefit from correlation discovery to map the outcome of the previous stages to required capabilities of the prototype. This can be enhanced by correlating the current project scope to comparable previous projects. Even DL with ANNs can play a role in the future to add self-learning for improved and increased targeting of prototype definition and scoping.

The *test* stage offers numerous opportunities to infuse AI: validating the user-centricity through correlation algorithms by comparing test results with data, metrics, and insight that are gathered and structured in the *empathize* and *define* stages, reducing the number of test cases through clustering and mapping to a required test coverage, discovering application or system usage patterns of functions and features, or parameter settings and other measures that enable a deeper understanding of product usage to influence iterative ideation stages.

In order for you to implement this AI design thinking model, you need to establish a stepwise approach that adheres to your particular DevOps maturity level and takes into account the specific project scope and goals. For instance, most organizations may get started with an AI design thinking model by focusing on the data preparation tasks and ML-based pattern and correlation discovery. Other model components, such as predictive analytics, text and voice, and DL aspects, may follow as subsequent steps.

# AI Design Thinking Value

The value of design thinking is widely recognized in the industry; the value of AI design thinking, however, is seen in principle, but still needs to be proven in the field.[11] As we have seen, all five stages of the design thinking process can gain from various AI capabilities, especially from the data preparation and ML capabilities. Deeper insight regarding the correlation and interdependency of the various process stages can be gained. Predictive analytics enables increased selectivity of alternative design ideas and also test cases.

More forward-thinking ideas, such as using image recognition, for instance, to analyze architecture overview diagrams, or using DL with ANNs to improve component interaction diagrams with self-learning means, have been deliberately left out of the discussion.

---

[11]See [13] for more information on the value and power of AI design thinking.

# Key Aspects of AI DevOps

In this section, we describe the key aspects and the value of AI DevOps. Similar to the previous section, we develop a simple AI DevOps model that maps the most noticeable AI features to the DevOps life cycle as depicted in Figure 7-2. To be consistent, we take the same AI capabilities (data preparation, ML, text and voice, and DL) that we have listed in the previous section. A short discussion on the value of AI DevOps concludes this section.

# AI DevOps Model

As we have pointed out earlier in this chapter, our primary interest is the intersection of development and operations, not the individual areas. In other words, we intend to detail out how AI improves the development stages and induces benefit for some or all operations stages, and vice versa. The goal is to simplify, accelerate, and automate the DevOps life cycle through AI. Similar to the AI design thinking model, we limit our discussion to an exemplary mapping exercise, thus leaving out other aspects of a more comprehensive framework or model.

Approaching this undertaking means for us to work out the most relevant DevOps intersection areas and map the AI capabilities from the previous section to those areas. The following list (although incomplete) gives the intersection areas, which we focus on. In doing so, we are leaving out some intermediate stages:

- **Operate and plan**: Operating applications, systems, and networks allows operations personnel and users to provide feedback to the planning team. The data to be collected (e.g., configuration and parameter settings, feature usage, operational patterns) can be explored, analyzed, and clustered and serves as input for ML modeling to predict user perception improvements.

- **Monitor and design**: Monitoring application and system behavior to identify prioritized improvement areas to be addressed during the design stage. Relevant data to be collected are logs (e.g., application, system, error, database logs), network utilization, memory and CPU utilization, and so on.

- **Implement and operate**: AI artifacts, such as ML models, can be implemented that optimize and simplify operations. For instance, non-trained ML models can be developed that use customer-specific data (e.g., application usage, SQL query execution) for training to accelerate workload execution and optimize application and system behavior.

As you see from the previous examples, the required data and the data flow play an essential role. This data either needs to be collected by operations (e.g., log or resource utilization data) and analyzed and consumed by development, or it needs to be prepared by development (e.g., non-trained ML models) and leveraged and integrated by operations.

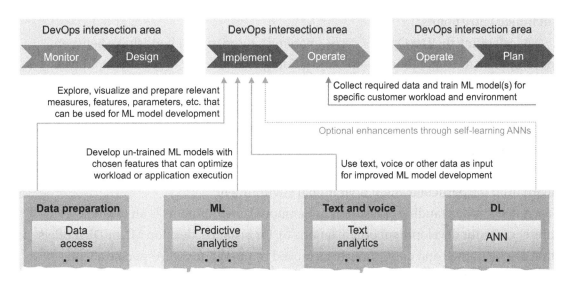

***Figure 7-6.*** *AI DevOps Model*

The AI DevOps model, as depicted in Figure 7-6, is a simplified abstraction, which describes exemplarily the mapping of the AI capabilities to the *implement-operate* DevOps intersection area. Not all AI capabilities need to be necessarily exploited for either intersection area; the art of applying AI to DevOps (and design thinking as well) is to define meaningful and relevant entry points that provide value to the organizations and then to grow over time. For instance, DL and ANNs may be implemented in a later phase.

The challenge is for vendors to hide the perceived complexity of AI through tools, RESTful APIs, and GUIs, which allows operations personnel to easily deploy and integrate AI DevOps enhancements.

## AI DevOps Value

The value of DevOps is widely recognized by many leading enterprises; widespread implementation, however, underpinned with corresponding tools and processes that truly and comprehensively help to bridge the gap between vendors and clients is still ongoing. AI-infused DevOps is an emerging field with its value more observed and noticed, rather than fully materialized and experienced. This will most likely stay this way for some time. As we have indicated earlier, AI DevOps is a journey with to-be-identified entry points.

Accelerating the DevOps life cycle, enabling a more fruitful handshake between development and operations teams, and gaining increased accuracy and rapidity of decision making are obviously the overarching value of AI for DevOps (AI DevOps). Detecting known problem signatures and patterns of operational issues can even trigger automated adjustments of the application or system environment, without manual intervention by operational staff or end users. Applying DL with ANNs has the potential to implement self-learning capabilities in the larger IT infrastructure that can adapt and learn from the particular client operational environment.

A deep understanding of customer operational environments is an obvious prerequisite for development organizations to, for instance, develop ML or DL models that can be trained and continuously improved by operations with specific client operational data.

## Key Takeaways

We conclude this chapter with a few key takeaways for design thinking and DevOps, as summarized in Table 7-4.

***Table 7-4.*** *Key Takeaways*

| # | Key Takeaway | High-Level Description |
|---|---|---|
| 1 | Design thinking and DevOps in place | A sound and mature design thinking and DevOps approach should be in place in your organization prior to infusing AI |
| 2 | AI design thinking value | Design thinking is a multidisciplinary approach, foremost about user-centricity, and democratizing participation; AI will improve these principles and make design thinking more relevant, accurate, and predictable |
| 3 | AI DevOps value | DevOps is about the process interleaving of development with operations; AI DevOps accelerates DevOps and increases accuracy and rapidity of decision making |
| 4 | AI DevOps scope | AI DevOps can be seen as the intersection of DevOps and AI; it is the enhancement of DevOps with AI |
| 5 | AIOps and AIDev scope | AIOps is looking at operations infused with AI without focusing on development; AIDev is looking at development infused with AI without focusing on operations |
| 6 | Identifying relevant entry points | AI design thinking and AI DevOps are a journey that require meaningful and consumable entry points |
| 7 | Data availability | AI design thinking and AI DevOps require data access, sharing, and preparation to gain relevant insight to significantly enhance both areas |

# References

[1]    Meinel, C., Leifer, L. *Design Thinking Research: Looking Further: Design Thinking Beyond Solution-Fixation.* ISBN-13: 978-3319970813, Springer, 2018.

[2]    Ney, S., Meinel, C. Putting Design Thinking to Work: How Large Organizations Can Embrace Messy Institutions to Tackle Wicked Problems. ISBN-13: 978-3030196080, Springer, 2019.

[3]   Simon, H.A. *The Sciences of the Artificial.* ISBN-13: 978-0262691918, MIT Press, 1996.

[4]   Hobcraft, P. *HYPE Innovation Blog: An Introduction to Design Thinking.* https://blog.hypeinnovation.com/an-introduction-to-design-thinking-for-innovation-managers (accessed September 27, 2019).

[5]   Kim, G., Willis, J., Debois, P., Humble, J. *The DevOPS Handbook: How to Create World-Class Agility, Reliability, and Security in Technology Organizations.* SBN-13: 978-1942788003, IT Revolution Press, 2016.

[6]   Forsgren Phd, F., Humble, J., Kim, G. *Accelerate: The Science of Lean Software and Devops: Building and Scaling High Performing Technology Organizations.* ISBN-13: 978-1942788331, IT Revolution Press, 2018.

[7]   Costello, K. *Gartner: The Secret to DevOps Success.* www.gartner.com/smarterwithgartner/the-secret-to-devops-success/ (accessed September 28, 2019).

[8]   IBM. DevOps for accelerating the enterprise application lifecycle. www.ibm.com/cloud/garage/architectures/devOpsArchitecture/0_1 (accessed September 30, 2019).

[9]   IBM. *Why apply design thinking to artificial intelligence?* www.ibm.com/design/thinking/page/badges/ai (accessed September 30, 2019).

[10]  IBM. *IBM Watson White Paper. Putting AI to work.* www.ibm.com/downloads/cas/JXRGQBVL (accessed September 30, 2019).

[11]  Volk, T. EMA. *Artificial Intelligence and Machine Learning for optimized DevOps, IT Operations, and Business.* https://bluemedora.com/wp-content/uploads/2018/10/EMA-BlueMedora-Top3-AI-2018-DecisionGuide-chapter-1.pdf (accessed September 30, 2019).

[12]    Sedig, K., Parsons, P. *Design of Visualizations for Human-Information Interaction: A Pattern-Based Framework (Synthesis Lectures on Visualization)*. ISBN-13: 978-1627057479, Morgan & Claypool Publishers, 2016.

[13]    Schmarzo, B. *Dell Technologies. Design Thinking: Future-proof Yourself from AI*. `https://infocus.dellemc.com/william_schmarzo/design-thinking-future-proof-yourself-from-ai/` (accessed October 4, 2019).

# PART III

# AI in Context

# CHAPTER 8

# AI and Governance

While AI is already widely leveraged in a vast set of use cases, we will see it become increasingly more commonplace in all industry verticals and affecting our societies. Inference of predictive and ML-driven insight into business processes can be characterized by a great deal of autonomous decision making, which may be perceived by some users as incomprehensible or elusive. Since AI-based decision making ought to be meaningful and human comprehensible, AI comes with a new dimension of governance imperatives designed to ensure transparency, trust, and accountability, taking into account explainability, fairness, and trackability.

Governance in the context of AI needs to go beyond the traditional existing data and information governance scope. One example is *fairness* as listed earlier, which is an aspect of AI governance that is related to ML and DL models as new AI artifacts that did not exist in conventional data-centric scenarios.

This chapter introduces data governance and shows how AI capabilities have been embedded to make data governance more efficient and more accurate. In this context, we apply AI on data governance. The second and even more important dimension though is how to govern AI which raises the topic of AI governance. We will explore the technical aspects and examine the various facets of AI governance, including to what degree existing data and information governance need to be extended, empowered, and upgraded to cater the needs for AI. We discuss specific challenges for AI governance, key regulations that are driving the need for AI governance, and ethical aspects for a trustworthy AI governance.

We explore some of the key aspects and technical capabilities of AI governance, neither doing this in a comprehensive way nor claiming completeness. We will explain the following key aspects of AI governance:

- **Rules and policies**: We show which enhancements for rules and policies are needed for AI-related methods and activities.

- **Glossaries**: Glossary content such as terms need to be broadened and added to cover all AI relevant aspects and definitions.

© Eberhard Hechler, Martin Oberhofer, Thomas Schaeck 2020
E. Hechler et al., *Deploying AI in the Enterprise*, https://doi.org/10.1007/978-1-4842-6206-1_8

- **Search and discovery**: ML models are costly to develop. To foster reuse, the need of models to be searchable and discoverable becomes essential. The same applies for many other AI artifacts.

- **Classification**: Classification and taxonomies need to be adjusted to cover and incorporate AI assets.

- **Provenance and lineage**: For trust in AI, the full life cycle from the origin of a model like on which data was it trained, who trained the model, the version history of the model, as well as all other aspects of AI artifact must be understood. Provenance and lineage are key capabilities needed here.

One of the key objectives of this chapter is to work out the specific and novel challenges that drive the need for AI governance and the characteristics that make up AI governance.

This chapter will introduce a few AI governance-related vendor offerings and principles, such as the IBM Information Governance Catalog, the IBM Watson Knowledge Catalog, and the IBM Watson OpenScale. A couple of sample use cases and key takeaways will conclude this chapter.

# Scope of Governance

Before we dive further into AI governance, we would like to elaborate on the scope of governance in general and how it relates to risk and compliance. In doing so, we provide a short review of governance, particularly in regard to data and information governance.

## Governance – A Short Review

In recent years, governance has evolved into a more holistic approach that also addresses the need for risk and compliance. Governance, risk, and compliance (GRC) pays tribute to manage technical and business risks across all organizations and at the enterprise level and to manage ongoing regulations and keeping up with the accelerating rate of regulatory and technology change, meaning to adequately react and accelerate consideration to the increasing number of new general and industry-specific regulations.

When we discuss *governance* throughout this chapter, we actually refer to the broader GRC scope.

Figure 8-1 is a depiction of these three terms[1].

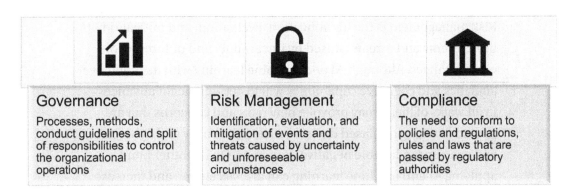

| Governance | Risk Management | Compliance |
|---|---|---|
| Processes, methods, conduct guidelines and split of responsibilities to control the organizational operations | Identification, evaluation, and mitigation of events and threats caused by uncertainty and unforeseeable circumstances | The need to conform to policies and regulations, rules and laws that are passed by regulatory authorities |

***Figure 8-1.***  *Governance, Risk, and Compliance (GRC)*

The following is a high-level description of these terms:

- **Governance**: The term governance[2] is usually associated with and even limited to corporate governance, which describes the processes, methods, conduct guidelines, and split of rights and responsibilities to control enterprise operations across all organizations. In its long history, corporate governance has evolved significantly to entail, for instance, data and information governance. Rapidly evolving advancements in technology and its impact on society are creating new challenges for corporate governance. Especially, intriguing use case scenarios that are enabled through AI characterized by autonomous decision making or driven and led by data scientists, such as self-driving vehicles, cognitive processing with voice and face recognition capabilities, image recognition leveraged for clinical judgment, or diagnosis, will again impose new challenges for governance to factor in ethical and legal aspects and the society as a whole. We will elaborate on this in the section "*Governance in the Context of AI.*"

- **Risk**: There are many different categories of events that are associated with risk, such as uncertainty of planned events achieving desired outcomes due to internal or external factors, or uncertainty in

---

[1]See [1] for more information on GRC.

[2]Governance may mean different things to different people; for the sake of the scope of this book, however, we stick to the corporate aspect of it.

financial markets, or cyber risks, project risks, IT unplanned outages and operational risks, and even natural disasters, to name just a few. Risk management is the identification, evaluation, and mitigation of those events and threats caused by uncertainty and unforeseeable circumstances. Although AI and machine learning with its predictive analytics and cognitive capabilities will foster significant advances in all walks of life, it may provoke new risks and concerns that are related to wrong or biased decision making, security and cyber risks, and even unpredictable negative outcomes. On the other hand, applying AI and machine learning can also accelerate and increase the accuracy of risk identification, understanding, and mitigation.

- **Compliance**: Regulatory compliance addresses the need to conform to policies and regulations, rules, and even laws that are passed by regulatory authorities or the government. It requires to establish methods and processes to achieve operational transparency, protection of critical data (e.g., personal health-related data), and adherence to often industry-specific business standards. Some examples are the EU General Data Protection Regulation (GDPR)[3] to guarantee more control to the owner of the personal data and giving them right to opt out, the Health Insurance Portability and Accountability Act of 1996 (HIPAA)[4] to modernize the flow of healthcare information and to protect vital patient data, or the Sarbanes-Oxley Act (SOX)[5] to protect the public from fraudulent or erroneous practices by corporations and other business entities.

GRC supports a holistic, integrated, and enterprise-wide view regarding all relevant aspects, such as processes, methodologies, strategic enterprise and organizational goals, relevant technologies, and roles and responsibilities of key stakeholders from outside and within the organization.

---

[3]GDPR addresses several important areas, such as lawful basis and transparency of your data processing activities, data security, accountability and governance, and privacy rights. See [2] for more information on the EU General Data Protection Regulation (GDPR).

[4]See [3] for more information on the Health Insurance Portability and Accountability Act (HIPAA).

[5]See [4] for more information on the Sarbanes-Oxley Act (SOX).

# Data and Information Governance

Data governance[6] was in its early days a cumbersome topic for data professionals who had to convince business and IT leaders why it is a critical capability for an enterprise and worthwhile investing in. Luckily, for data professionals in many countries, in 2020 this problem is mostly gone. Today, usually there are three main drivers for data governance: process efficiency, regulatory compliance, and improved customer experience. By setting standards around data management, business and IT processes are streamlined and cost is reduced. A never-ending flood of regulations around the world force enterprises to manage information assets in compliance with these regulations which is another motivation for data governance. Examples[7] of regulations include:

- General Data Protection Regulation (GDPR) in Europe

- California Consumer Privacy Act (CCPA) in the USA

- Protection of Personal Information Act (POPI) in South Africa

- Basel Committee on Banking Supervision Standard 239 (BCBS 239)

- Solvency II in insurance

- MiFID II in financial markets

- Health Insurance Portability and Accountability Act (HIPAA) in the USA

Industry-leading enterprises often use data governance to drive improvements in data management aimed at enhancing the customer experience. Data governance is a key component of GRC to ensure that relevant data to be used within the GRC and organizational business context adheres to required data quality service levels and is trustworthy, complete, and consistent.

We only introduce a few key concepts of data governance to set the stage for our discussion[8]. A data governance program is comprised of three constituents: people (and their organization), process, and technology. Figure 8-2 shows a conceptual overview

---

[6]See [5] for more information on data governance.

[7]Details on these regulations can be found in the following references: [6], [7], [8], [9], [10], and [11].

[8]If you want to learn about data governance in depth, these books provide excellent coverage of many critical aspects of data governance: [12], [13], [14], [15], [16], [17], and [18].

of data governance. There are variations of data governance[9] definitions as well as the disciplines and capabilities included which are beyond this introduction for comparison and analysis. Figure 8-2 shows the essential parts from our perspective. Let us start at the top with the business outcomes of data governance.

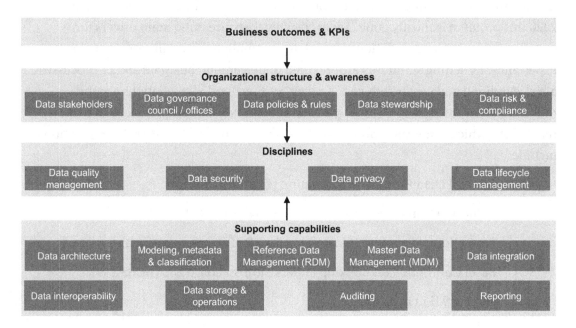

***Figure 8-2.*** *Data Governance Overview*

To run a data governance program successfully, it has to be set up with well-defined key performance indicators (KPIs) articulating the business objectives and a control mechanism to measure the progress toward these goals. Each KPI needs accountable owner(s) as well. Setting goals and trying to measure progress toward them require well-defined processes. Efficiency is gained if these processes are automated as much as possible using technology. Here are some examples of beneficial business outcomes achieved with data governance in the areas of process efficiency, compliance, and improved customer experience:

---

[9]DAMA International is a non-profit, vendor-independent global association of technical and business professionals dedicated to advancing the concepts and practices of information and data management. The Data Governance Institute is also a vendor-neutral institute but focused only on the topic of data governance. Both of them have definitions of data governance – but they are not identical. You can find more on both of their websites located here [19] and here [20].

- *Clear and transparent communication of the goals* of data governance is streamlined through standardized data management practices and processes. Data governance councils or data governance offices with clearly defined roles, decision rights, accountability, and control mechanism facilitate the definition and execution of these standardized data management practices and work with business and IT departments to get them properly implemented.

- *Standardized, consistent, and uniform data management practices* are a prerequisite for enterprise-wide data-centric initiatives such as MDM, data lakes, and so on.

- *Agility and scalability of the IT operations* on the organizational, business, and technical level are significantly increased through clear rules on how to define, change, and implement processes affecting data. The scale aspect of managing data is becoming increasingly important as the data volumes grow at an accelerating speed.

- *A governed approach to managing data* allows reuse of data processes and information assets avoiding duplication and redundancy. As the data governance program matures over time in organizations, more and more redundant data silos and inconsistent data processes are identified and can be consolidated reducing IT costs and the operational risk (less systems to secure).

- *Business decision makers benefit from data which they can trust* because it is managed with the right (certified) data quality, and they know its origin through data lineage. At any time, business decision makers can also review the entire data process documentation around the information asset they are using, thus getting full transparency into every aspect of the data they are using.

- *Effective data governance programs establish data privacy policies and standardize tool* sets to identify information assets relevant for regulations affecting data privacy. Furthermore, consistent processes and tools are implemented to address requirements from data privacy regulations like GDPR around consent management, data subject access rights (e.g., right to review, right to be forgotten, etc.).

Such standardization through data governance reduces IT costs for compliance and mitigates the financial risks in terms of fines for noncompliance.

- *Many regulations such as GDPR require to manage certain data assets* like personal identifiable information with appropriate security controls. Common standards for data security at rest and in motion, as well as access controls, simplify the implementation of such required controls and reduce costs.

- *Certain regulations and laws also indicate retention times for data* affecting the data life cycle management. Setting standards through data governance is beneficial for multiple reasons. First, it ensures data is not being deleted before it is permitted from a regulatory angle reducing the risk for noncompliance. Second, by establishing practices to delete data once, there is no business or regulatory retention requirement anymore, which reduces storage costs and also mitigates risk because it can't be compromised anymore.

- *Retaining a customer is easier* compared to getting a new customer or winning a lost customer back. Another and even more critical benefit is that integrated, instead of duplicated, customer records avoid angering customers by sending them, for example, marketing materials multiple times. Data governance programs are already used by enterprises to analyze and improve data practices in business processes where the customer interacts with the enterprise. For example, if duplicate customer records across multiple systems cause that the same customer gets the same marketing material multiple times, this customer might get annoyed and does business elsewhere in the future.

Since each company implements a specific scope of data governance fitting their enterprise, other business benefits and outcomes are possible as well.

The implementation of a data governance program requires an organizational structure and people creating awareness around the data governance program and driving it from a day-to-day perspective. From an organizational perspective, different terms have been used to describe the data governance organization structure, such as data governance council, data governance offices, chief data office, and others. The name in our experience is less important than the fact that organizational units exist which are responsible to drive data governance across the enterprise. We use the term data governance office in this chapter. Typical roles either within the data governance office or as key stakeholder relevant for the data governance office are as follows (this list is not exhaustive):

- **Data stakeholder**: A business executive responsible for a business process, which contains data entities that need to be governed. Examples include executives of departments such as sales, marketing, supplier, HR, and so on since many of their business processes refer to regulated data such as personal data.

- **Chief data officer** (CDO): A chief data officer is usually a senior executive with a strong business background. A CDO is responsible for defining data strategies supporting and enhancing key business initiatives. This usually includes strategies for data architecture, adoption of AI, and data governance across the enterprise.

- **Data protection officer** (DPO): This is a role which enterprises must have according to article 37 of the GDPR regulation if the enterprise processes EU citizen's data and certain critical criteria are met. A DPO serves in a senior security leadership role overseeing the implementation of the necessary data protection measures for data to comply with GDPR requirements. They serve as point of contact for the company to GDPR supervisory authorities. Article 39 of GPDR outlines the DPO responsibilities which include education and training of employees on GDPR compliance, execution of audits to ensure compliance, and maintaining complete records of all data processing activities including the purpose of such processing activities.

- **Chief information security officer** (CISO): The CISO role must bridge the gap between disciplines like IT security and business. With a very good understanding of the business goals, a CISO must define a strategy for information security to reach appropriate levels of data security without limiting the agile nature of modern business processes. Key areas include cyber risk and intelligence management, security operations, protection against data loss and fraud, identity, and access management (IAM) and audit processes for the entire IT security with accountability to the board of directors.

- **Chief compliance officer** (CCO): The CCO has to make sure that the company conducts its business in full compliance with all national and international laws, regulations, professional standards, accepted business practices, and internal standards.

- **Data steward**: A data steward (or team of data stewards) is accountable for one or more data domains (e.g., employee data, customer data, etc.). The data steward makes sure that all attributes of the data domain are well defined and described with business terms to concisely capture the semantical meaning. The data steward curates metadata, policies, and rules around the data domain and manages the day-to-day resolution of data quality issues in the owned data domain.

- **Enterprise information architect/information architect**: Enterprise information architecture[10] is a discipline within enterprise architecture[11]. The enterprise information architect or information architect is responsible for defining the data architecture in support of the business architecture. The data architecture covers all architecture aspects from metadata management, reference data management, MDM, transactional data processing, and of course the end-to-end considerations for all analytics systems.

Some of the roles will always be part of the data governance office such as the data stewards or frequently part of it, like the DPO, while other roles provide key requirements toward the data governance office. An example is the CISO role in the area of data protection.

As you already noticed, data governance is a broad theme and hence a data governance office divides the implementation of data governance in a set of four core disciplines:

- **Data quality management**: Data quality management is the set of processes and tools used to manage data quality in the context of a given business scenario. It can be measured in many different dimensions such as completeness, format and domain compliance,

---

[10]The following sources provide comprehensive coverage on the topic of enterprise architecture and enterprise information architecture: [21], [22], and [23].

[11]Chapter 4, "*AI Information Architecture,*" provides more context on how information architecture got expanded for AI and fits into the enterprise architecture realm.

standardization, verification against trusted sources (e.g., postal address verification), and others. Data quality policies articulate the business requirements, and data quality rules define the technical specification corresponding on the data quality policies. Data quality profiling and monitoring tools are used to assess if data assets are compliant with the data quality policies. If data quality exceptions are found, tasks for the data stewards are created. With modern data governance tools, consistent and repeatable data remediation processes are executed by the data stewards, fixing data quality issues.

- **Data security**: Data security[12] addresses all aspects of protecting data against unauthorized access from external or internal attackers. Data security covers topics such as authentication standards and single sign-on (SSO), encryption of data at rest and in motion, data masking standards for sensitive data for development and testing purposes, processes and tools for secure data erasure, secure engineering standards for all applications used in IT, and deployment of monitoring tools recording all system access and data changes.

- **Data privacy**: Data privacy concerns the right of an individual that personal information is properly handled and stored, meaning that personal information is strictly used appropriately protected within the business process for which it was collected. GDPR and other privacy regulations mandate that otherwise consent has to be given by the individual for additional and well-defined use triggering the need for consent management capabilities. Such privacy regulations also grant data subject access rights to individuals like the right to review or the right to be forgotten. Data security and data privacy are not the same: for dealing with certain aspects of data privacy, you need data security to be able to protect personal information through encryption. However, you can implement data security without having data privacy capabilities such as consent management deployed.

---

[12]The ISO standards 27000, 27001, and 27002 are relevant standards for data security. You can find them here: [24], [25], and [26]. Cloud service providers are often asked if their cloud services are compliant with these standards.

- **Data life cycle management**: Data life cycle management considers the life cycle of a data from the initial creation to the final erasure. Certain data assets must be kept by an enterprise for a certain number of years as required by retention mandates of regulations. The opposite exists as well: data has to be deleted once the business relationship ended. With data volumes growing at an increasing speed, being able to delete data when it is no longer required is a critical need to control storage costs. Any data kept longer than needed is also a liability because attackers can compromise it. Hence, as part of data governance programs, appropriate retention policies for data are defined and processes implemented to manage the data assets in compliance with the retention policies. Another aspect of data life cycle management is the frequency of data accessed. Fast storage is much more expensive compared to slower, but cheaper storage. Usually, if data is getting older, it is less frequently accessed. Policies can be defined when data can (and should) be offloaded from faster to slower storage systems optimizing storage costs. This is a common technique controlling storage costs in data warehouses using SQL federation running across systems with different storage speeds.

In support of these four major disciplines, there are a number of enabling capabilities as you saw previously in Figure 8-2. These capabilities are usually supported with specialized technology, providing optimized functionality for the particular capability area. We provide a short summary of each:

- **Data architecture**: The enterprise information architect or information architect in your enterprise (or possibly a team of them if your enterprise is large) defines the data architecture in support of the business architecture. The data architecture is not a single system due to various different needs of your business for transactional and analytical systems. Data architecture focuses to deliver scalable and cost-effective systems avoiding data redundancy wherever possible. A well-defined data architecture anticipates changes required by future business needs and allows rapid change if required. Tools[13]

---

[13]You can learn more about these tools here: [27], [28], [29], and [30].

from Software AG, LeanIX, and Sparx Systems or open source like ArchiMate help with the modeling of relevant architecture artifacts.

- **Modeling, metadata[14], and classification**: Data modeling tools are used to model the metadata structure of logical and physical data models describing the structure of data entities as well as the relationships between them which are needed for the business of an enterprise. Logical and physical data models are just two examples of technical metadata, and other technical metadata artifacts are used to describe the structure of business reports, data transformation jobs, and so on. Metadata management tools, better known today as data governance catalogs, are not only used to manage technical metadata; they are also used to manage business metadata such as business terms, a broad range of policies such as data quality and data privacy policies, retention or data access policies, and data rules. Ontologies are used to provide taxonomies for business terms and data classes defining the semantical meaning of attributes in logical data models. Semantical classification algorithms, applied while discovering new data sources, map the attributes of the discovered source data attributes to existing data classes and business terms. Another type of metadata is operational metadata, which comes in many different flavors, addressing operational statistics when data quality reports were run or tags and user comments associated with information assets in the catalog, indicating if a particular information asset was useful for a particular task. Another critical set of capabilities for a data governance catalog is the business lineage and data lineage functionality to understand the heritage and relationships of data.

- **Reference data management (RDM)**:[15] This capability addresses the need for enterprises to manage reference data. Reference data examples are lists of disease codes in healthcare, a country code list,

---

[14]Managing business metadata is well described in this book [31], and metadata management with a broader perspective is discussed here [32].

[15]If you want to dig deeper into reference data management, we recommend the following resources: [33], [34], and [35].

or a list of different permissible account types. Inconsistent reference data is a severe data quality problem. For example, it will be very difficult to correctly report revenue by country in an enterprise DWH if the transactions for various transactional systems feeding the DWH used different country codes.[16] Managing reference data consistently is the reason why companies started to implement dedicated RDM solutions. In recent years, there is a trend that RDM capabilities become part of the data governance catalogs[17].

- **MDM**: This capability provides efficient master data management functionality and we will cover it later in this Chapter and Chapter 9, *"Applying AI to master data management"*.

- **Data integration**: There are several different techniques for data integration which are extract-transform-load (ETL), data replication, data virtualization, data streaming, and real-time APIs. ETL is basically batch processing of large amounts of data and is often used for initial loads from source to a new system, or batch feeds into a DWH or data lakes. As part of the batch processing complex, data cleansing and data transformation routines are often applied. Data replication is a near real-time movement capability of data changes from one system to one or multiple other systems. Use cases include trickle feeds into analytics systems which require more real-time data input or keeping a standby system for recovery in sync with the primary system. Data virtualization (previously known as data federation) basically hides the complexity of multiple data sources with possibly even different SQL dialects[18] from application developers. Using data virtualization, multiple sources appear as a single source to the application developer, who can focus on the application function, and the data virtualization layer handles the execution of the SQL query under the covers against multiple sources. One of the advantages

---

[16]The need for reference data management for big data and business intelligence solution is discussed in depth in a research paper located here [36].

[17]The data governance catalog solution from Collibra and the Watson Knowledge Catalog from IBM are examples of this trend. You can find more about them here [37] and [38].

[18]Database vendors like Microsoft, Oracle, SAP, or IBM all have some vendor-specific nuances around their SQL implementation.

of data virtualization over ETL and data replication is that there is no copy of data created, saving storage costs. Data streaming is real-time data movement possibly with analytics built in, which is particularly effective in Internet of Things (IoT) scenarios where very large data volumes are created in real time. Data streaming can be used to analyze in real time if there are valuable data elements in the data stream which should be persisted or further processed while many other data elements are immediately dropped (and never even written to disk). Real-time APIs provide transactional, scalable, and high-performance data access, which is often required by system of engagement applications like mobile or online channels. The data integration capabilities should expose all metadata describing the data structures and, if applicable, all transformations into the data governance catalog infrastructure enabling full data lineage.

- **Data interoperability**: Data interoperability[19] is becoming increasingly important since an increasing number of enterprises recognize the beneficial effects of value ecosystems, where companies collaborate with one another to deliver innovative solutions to their clients through partnerships. Another driver is the adoption of hybrid cloud strategies where a portion of the IT capabilities are in the enterprise data center and another portion is consumed from one or multiple public cloud vendors. Both drivers bring data interoperability challenges – previously existing just between systems in the corporate data center – even more into the spotlight. With this increased focus, companies as part of their data governance programs establish standards for data interoperability. These standards trigger contracts between systems which create and exchange data in such a way that the semantical meaning and scope of the exchanged data is well defined.

- **Data storage and operations**: The storage price per GB gets constantly cheaper, and with cloud object storage available on public clouds, there is even a promise of more or less unlimited storage

---

[19]The Data Interoperability Standards Consortium is focused on this particular topic. You can learn more about that interoperability on their website located here [39].

capacity for an enterprise. Although data storage should not be a significant problem anymore, the explosion of data volumes in recent years, and its prediction for the near future, does offset the price reductions for storage. Putting data onto cloud object storage or other persistency services on public clouds is cheap on the way in, but many cloud vendors have very expensive egress[20] charges once you intend to pull the data out to move it back to your own data center or another vendors cloud. Some data is frequently accessed, while other data is rarely accessed (e.g., a data backup) or only kept in case an audit or compliance case requires it. Hence, optimizing data placement onto different types of storage systems in a cost-aware way, providing different speed and throughput based on usage patterns, is another dimension to consider. From an operations perspective, making sure that data is securely erased from disks to avoid restoration with specialized tools once storage systems get replaced is another critical aspect to avoid incidental data breaches by dumping old storage systems.

- **Auditing**: Data governance programs require appropriate audit functionality deployed on all critical systems to have audit trails (who accessed and modified data and when). State-of-the-art audit tools[21] are capable of detecting unusual data access patterns by a user and can terminate access in real time preventing potential data breaches. For example, an unusual data access pattern could be the request to read all entries of the customer data table in a database by a database administrator. This could potentially indicate a disgruntled database administrator who wants to create a copy of the customer data before leaving the company.

- **Reporting**: As mentioned, data governance needs to deliver KPI-related business value. Without reports, ideally fully automated, the effectiveness and progress through the data governance program cannot be judged. Hence, appropriate reporting capabilities are an essential part of a successful data governance program.

---

[20]You can read more about this here [40].

[21]IBM Guardium is one such tool with this capability; see [41] for more information.

Enterprises today have a huge amount of data assets. It is usually impractical and impossible to implement data governance programs covering every single attribute of every single data asset an enterprise owns. The objective of a data governance program is to govern the relevant aspects of the data landscape where it provides value to the enterprise. For example, a pharmaceutical distribution company, which ships tens of thousands of packages from their distribution centers to pharmacies and hospitals each day, has a compelling business case to not only standardize but actually verify the postal addresses on customer records against a postal dictionary.

If a package comes back to the distribution center, it takes a person on average 10 minutes to correct the address, place a new label onto the package, and ship the package again. If you summarize the cost of labor and the second shipment, multiplying this by a few hundred to a few thousand cases per day (due to low data quality on address), you can easily see that this adds up to a significant cost. Hence, there is a valid business case to govern the data quality of address information with the highest-data-quality standards as effectively as possible, reducing the operational costs significantly and contributing to the net profit. Using this as an example, good data governance programs are not trying to boil the ocean; they rather focus their budget and efforts around the most relevant business outcomes.

# Infusing AI into Data Governance

Given the broad range of topics belonging to data governance – as you have seen in the previous section – this section focuses on using AI in two areas of data governance. First, we show how AI is applied to metadata and data quality management. Second, we explore AI usage in data security.

## AI Applied to Metadata and Data Quality Management

The explosion of data volumes is not only a challenge for the storage cost; it also leads to a growing number of data sources, which need to be inventoried into the data governance catalog. Similar to a catalog in a library where each book has a reference describing the book and where it can be found, every data source needs an entry in a data governance catalog. A non-registered data source cannot be found or governed effectively. Figure 8-3 shows some of the relevant areas a data source entry in a data governance catalog should have.

Unfortunately, quite a number of data lake initiatives failed because they were ungoverned, meaning that many sources get loaded daily without implementing an inventory description of each loaded source in the data governance catalog. After some time, these ungoverned data lake initiatives lack required knowledge in regard to which particular sources were loaded into the data lake and where they can be found. Data scientists had no authoritative records in terms of data sources available, which information it contains, its level of data quality, permission to use it, and so on. More often than not, this fuzziness of metadata knowledge eventually required the data lake initiative to start all over again. However, a restart only made sense if the approach to govern the data lake got more intelligent, which means the approach to metadata management had to be infused with AI.

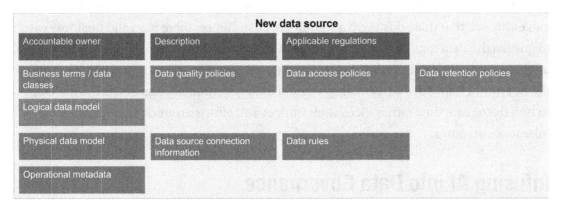

*Figure 8-3.* *A Conceptual Data Asset Card in a Data Governance Catalog*

Imagine you have a data source like a relational database with 50 data tables with 20 attributes on average, totaling to 1000 attributes. Assume you would be the data steward responsible to import all technical metadata, such as the logical and physical data model into the data governance catalog, manually assigning the applicable business terms and data classes, the applicable regulations, relevant policies, and so on and completing the full description of that data source in the data governance catalog.

To be able to make the correct assignments of the business terms, data classes, and relevant data quality policies, you also need to perform data profiling and semantical classification tasks. If this needs to be done often, it obviously involves a lot of human labor, leading to an unacceptable expensive endeavor. In recent years, the adoption of data lakes as a foundation for analytics environment to develop and train AI models has found many adopters.

The manual, not scalable management of metadata though is not the only driver for adoption of AI into metadata management. The preceding scenario assumes that business terms have been defined already in the data governance catalog. With a rapidly growing number of regulations, just having relevant business terms in the data governance catalog has turned into a barely manageable labor-intensive effort. Thus, another problem is the intelligent proposal of business terms related to regulations based on AI.

Dark data is basically any ungoverned data source in an enterprise, which causes many business problems. For example, if dark data is not needed, IT costs without benefits incur. If dark data is sensitive and an internal or external attacker accesses it, it can be a significant liability in reputational as well as financial risks. Personal identifiable information (PII) in dark data is problematic under many privacy regulations, causing potentially noncompliance issues. While many data governance programs made significant progress in reducing the amount of dark data in an enterprise for structured data sources by inventorying a large percentage of them into the data governance catalog, the unstructured data side is literally still in the dark. Examples of unstructured data sources in private and public cloud environments are content management systems, email systems, Microsoft Sharepoint, Box, and so on. This has several root causes.

First, data governance catalog vendors did not support unstructured sources in their metadata management tools for a long time, which has only improved in the last few years. Second, data profiling and semantical classification algorithms were for a while focused only on the structured data sources. Third, identifying PII in an unstructured data source is much more complicated compared to a structured data source.

If you have a column in a relational database and most of the values in that column seem to be last names, it is not that hard to conclude that this column is semantically a very likely candidate to be an instance of the data class *last name,* and the appropriate business term should be associated with that column. In a free form unstructured document, the value *Washington* could be a last name, the name of a city, or the name of a state.

Making a decision which semantical concept is applicable is a more complex problem to solve in unstructured data sources. Despite these obstacles, GDPR and CCPA are pushing companies to either delete or secure and properly manage personal data. With terabytes or petabytes of unstructured data, companies don't know where or how to start, and traditional approaches of indexing unstructured data on such volumes take months or years to complete. Companies also placed a lot of unstructured data into public cloud. API calls to index large sources on the public cloud are very costly. Thus, appropriate statistical sampling techniques combined with AI-infused capabilities for metadata discovery and classification of unstructured data sources are needed.

Data quality management with techniques such as data profiling and data quality monitoring is a well-studied field in academic research[22]. There are many useful data quality functions in commercial and open source software tools available today. However, data quality management is hitting its limits due to the rapid growth in number and size, as well as the broadening variety of data sources (very similar reasons why traditional approaches to metadata management are now insufficient). For example, your data profiling tool might have a data rule to discover whether or not a particular column is a US social security number by checking if the values adhere to the format XXX-XX-XXXX, where X represents digits between 0 and 9, and valid patterns adhere to certain rules.

With a broader variety of data sources, the likelihood that your data profiling tool of choice has all the required data rules for data quality out of the box is decreasing. Hence, the question arises, if you can use AI to discover new data rules based on ML and suggest them to the users. Similar to the discovery problem on unstructured data, the very large data volumes today force more and more adoption of appropriate statistical sampling techniques for data profiling and monitoring to keep the runtimes of such tasks acceptable (and this assumes already elastically horizontally scalable implementations of data profiling and monitoring on platforms like Apache Spark).

Discovering data quality issues with data profiling is a first step. Afterward, you often need to remediate and correct the data quality issues which was another source of tasks for data stewards to manually resolve. The problem of manually fixing data quality issues also reached a point where just using manual approaches is no longer sufficient. We will study examples where AI techniques are used to identify and correct data quality issues in Chapter 9, "*Applying AI to master data management.*" An even better approach instead of using AI to correct data quality issues is to leverage AI to prevent data quality issues on data entry.[23]

After we have gained an understanding of some of the limitations in metadata and data quality management, let us explore some of the AI capabilities which have been infused into data governance catalog and data quality management technology

---

[22]This academic research paper provides a comprehensive overview on data quality assessment methods and you can find it here [42].

[23]Bosch won the CDQ Good Practice Award in 2018 by using AI on data entry preventing data quality issues in a certain data area. You can find more details here [43].

already. The first AI-infused capability we explore is leveraging AI to extract and propose new business terms for new regulations as candidates into a data governance catalog automatically.

Figure 8-4 shows the conceptual flow of this new capability. Let us take a closer at the four stages of this flow.

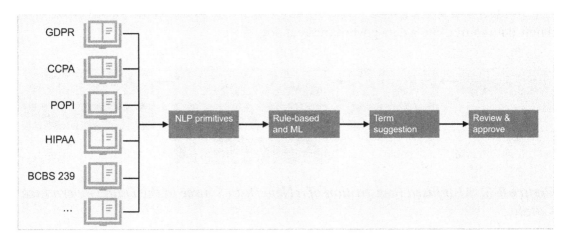

***Figure 8-4.*** *Using AI to Suggest New Business Terms from Regulations*

1. *NLP primitives* for sentence boundary identification, tokenization, lemmatization, parts of speech, dependency parsing, and expanded shallow semantic parsing are used to process the regulatory input text.

2. *Rule-based and ML* techniques are the next processing step. A rule-based extraction engine using the Annotation Query Language (AQL) which sits on top of the NLP primitives can extract the relevant features for ML models. Entity extraction with ML algorithms such as conditional random fields or ANNs, relation extraction with maximum entropy or ANNs, and many more techniques can be applied in this step.

3. *Term suggestion* for one or multiple new business terms can be derived.

4. *Review and approve* allows a data steward to review the proposed business terms, optionally make corrections, and then decide for each business term if it gets approved as a new term in the glossary of business terms within the data governance catalog or if it is rejected.

Now take a look at Figure 8-5 which is showing the AI-infused registration process of a new data source into a data governance catalog.

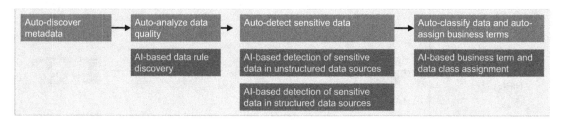

***Figure 8-5.*** *AI-Infused Registration of a New Data Source in the Data Governance Catalog*

A previously lengthy and time-consuming series of manual steps has now been automated to mostly a review and where necessary only manual correction steps:

1. Once the source connection is established, all metadata is automatically discovered and ingested into the data governance catalog.

2. The data quality assessment is automatically triggered on all discovered data attributes. For attributes where the discovered pattern distribution suggests that the values adhere to some conforming overarching pattern based on AI, a new data rule is predicted and suggested to the data steward (in case no matching data rule was found in the data rule library). The data steward is also provided with a percentage of how many values would be compliant with this rule. If the data steward agrees that the suggested rule is accurate, the data rule is added to the data rule library.

3. Given the differences of structured, semi-structured, and unstructured data sources, tailored techniques need to be applied to determine whether or not a data source contains PII. Random forest classifiers for accuracy and reasoning, predictive sampling

This as background suggests three key questions. First, more sophisticated attacks? Yes, they do. For example, A⁻ password crackers. Second, are AI models or the data they ⌐ a potential attack? The answer is yes, as attackers steal AI mode⌐ privacy by trying to steal training data which might contain PII data ⌐ inversion attacks. Third, can AI be used to protect you against security an⌐ threats in particular? Again, the answer is yes, and Table 8-1 shows you some r⌐ areas where AI-powered solutions are available today. The idea of Table 8-1 is to sho⌐ you where in security, and in particular data security, novel AI-based solutions might be able to effectively[28] help you.

Software vendors[29] have offerings for the preceding areas available today. As part of your data governance program, particularly related to data security, you should thus look into what is needed to secure your data assets appropriately, which might require the use of several AI-infused software capabilities in the security area.

With a basic understanding of data governance, we can now look at the impact of how AI can be governed with the capabilities from the data governance space. We will also show which additional capabilities will be required in the data governance domain to extend it so that it can be effectively used in the AI domain as well.

# Governance in the Context of AI

Governance in the context of AI can be very pervasive and broad, reaching far beyond the pure technical aspects, which are just a means to get it right; governance in its roots are policies, standards, and compliance aspects that an organization should uphold.

In this section, we show the challenges for AI governance and how traditional information governance will be impacted by AI and machine learning. In addition, we briefly elaborate on some regulations driving AI governance for a trustworthy AI.

---

[28]AI is not solving everything in IT or data security in particular. There are many critical voices such as [48] indicating that you need to carefully assess the capabilities provided by software vendors to see where they provide reliable value adds and where you still need to rely on your security team expertise.

[29]For example, IBM offers IBM Guardium, IBM QRadar Advisor with Watson, IBM QRadar User Behavior Analytics, IBM Resilient, IBM Trusteer, and IBM MaaS360 with Watson to address all these areas with AI-infused capabilities.

---

using clustering techniques, and entity extraction using ANNs are just a few examples of AI capabilities which could be applied in this step to determine whether or not a data source contains PII[24]. Leveraging such AI capabilities under the covers in this step, the data steward gets an assessment result whether or not the scanned source contains PII. The data steward can review the assessment result and decide whether or not the result was accurate by checking the prediction confidence or by reviewing relevant entries if necessary. In case the PII detection result was wrong, the data steward can correct the assessment.

4. In the next step, data classes representing semantical concepts and business terms are assigned to the discovered attributes in a data source based on AI. The attribute metadata, the profiling results from the previous steps, and the data values are considered as input. For example, if a column based on semantical classification was detected as containing values representing cities, the data class representing the semantical concept of *city* would be assigned to that column, including the business term *city* describing with business metadata what a *city* is and how and why this attribute is used. Again, the data steward can review if the proposed assignments are accurate and approve or override them. For AI-based data class and business terms assignment, a data governance catalog needs to provide a different set of AI techniques to deal with the following cases effectively:

   a. In case no labeled data is available, unsupervised learning techniques can be applied. This case applies when there are no data assets yet where attributes have assigned business terms.

   b. Different application packages (e.g., SAP, Salesforce, etc.) or different custom applications might use very different naming conventions. So even if the AI-based term assignment looks at metadata describing the attributes as well as the actual

---

[24]Using machine learning presents risks to data privacy in its own right since trained models might be vulnerable to de-anonymization attacks possibly disclosing PII. You can learn more about this here [44]

attribute values, a model which works in one instance might not work in another instance. In such a case, instance-based learning methods with appropriate similarity measures need to be applied for the AI-based term assignment.

c. A glossary of terms might have thousands or tens of thousands of terms. So even with labeled data, this might turn into a multi-class, multi-label classification problem very fast with maybe only 5–50 instances per class. To overcome this challenge, techniques like nearest neighbor or transforming the problem into a binary classification problem can be used.

For steps 2–4 in this process, active learning capabilities can be applied to continuously improve the results of the AI-based capabilities. The value proposition of this AI-infused, automated process is very significant. For example, in step 4, without the AI-based suggestions for data class and term assignments, a data steward would have a need to search for appropriate data classes and terms in the catalog before being able to establish the relationships between them and the attributes. Entering the keywords to search for data classes or terms, reviewing the result lists, selecting the applicable entry, and then creating the relationship to the attribute in question is time consuming. With the AI capabilities in place, the data steward in vast majority of the cases is done with a single mouse click approving the suggestion, taking out a lot of time and saving labor costs. While there is a lot more to discuss on this topic, we hope with this introduction you got a basic understanding why it is very valuable to apply AI techniques to metadata and data quality management. Let's now move to another data governance area where an infusion of AI capabilities was done in recent years which is data security in the next section.

## AI Applied to Data Security

Data security is a critical aspect of data governance. The 2019 Cost of Data Breach Report[25] indicates that the global average for a data breach in the USA is $3.9 million with an average cost per case of $8.19 million. The cost is typically divided into direct costs (fines for noncompliance with regulations, settlements, etc.) and indirect costs

___

[25]See [45] for the report, which contains a lot more details on the various dimensions of costs and others.

nal damage, etc.). If data breach costs are expensive, the obvious next question d to signs regarding a trend of increased data security issues. According to the -Force Threat Intelligence Index 2020,[26] the trend for 2020 and beyond shows an cipated growth of security-related incidents. Thus, the problem of potential data eaches is not going away anytime soon. The largest source of data-related breaches (approximately 50%) are external attackers. Security professionals are urgently needed to help prevent incidents or – if an incident occurs – identify and remediate the incident as quickly as possible. But how easy is it that enterprises can hire security professionals for their IT staff? According to this ISACA report on the State of Cybersecurity 2020,[27] 57% of the enterprises in their survey indicated to have open positions for security positions where 30% are open for more than 3 months and 29% are open for more than 6 months until suitable candidates were found.

*Table 8-1.* *Security Areas Infused with AI*

| Security Area | Key Value Points of AI |
| --- | --- |
| Data protection / security | Improved protection of data detecting and preventing illegal access to data |
| Security operation center | Prioritize alerts, classify indicents according to MITRE classification, collect relevant information and automatically recommend best action for the security analyst |
| Digital identity and fraud detection | With the help of AI, identify and prevent attacks on digital identies or fraud |
| Endpoint management | Leverage AI to protect your endpoints more effectively against attacks |
| User behavior | Leverage AI to detect anomalies in user behavior indicating potential threats and act on them |
| Intelligent incident response | Use AI for intelligent decision making in selecting and orchestrating the right incident responses |

___

[26]This report can be found here [46] and provides a lot of useful details on the various categories of attack, a breakdown in internal vs. external threats, and so on.

[27]The report containing the cited details can be found here [47].

# Beyond Traditional Information Governance

In this section, we elaborate on the mutual interference and impact of information governance and AI. Under the influence of AI, information governance needs to become broader by including new AI-related artifacts as well as AI-related challenges, extensions within the scope of governance. As we have seen in the previous sections, this causes challenges for application and enactment of AI governance. On the other hand, AI increases confidence in the data quality dimension of information governance by simply leveraging AI and machine learning algorithms and techniques, for instance, by increasing trustworthiness and accuracy for data matching tasks as you will see in Chapter 9, "*Applying AI to Data Governance and MDM.*"

AI represents a significant opportunity to advance some information governance aspects. For instance, AI accelerates insight and reduces the circle of action; after the corresponding action has been performed, it also reduces the time to realize the impact and revert or persist the change. AI furthermore helps in answering the questions fast instead of a human trying to read through a whole database.

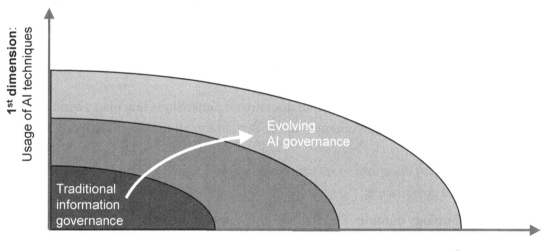

***Figure 8-6.*** *Interplay of Information Governance and AI*

Figure 8-6 illustrates the two-dimensional characteristic for traditional information governance to develop into AI governance:

- **First dimension**: Growth of information governance by leveraging AI and ML techniques. In this dimension, AI and ML constitute an improvement of information governance. As previously shown in this chapter, ML can significantly reduce the amount of human labor in source analysis and classification. Another example is to apply ML methods to identify operational risk patterns.

- **Second dimension**: Growth of information governance driven by the need to govern additional AI and ML artifacts, scenarios, terms and definitions, rules, and policies.

Without making the picture overly complicated, we could even add a third dimension to take into consideration evolving organizations climbing up their ladder to AI, offering new opportunities and new challenges, which need to be addressed and entertained by the AI governance framework. This evolving AI governance scope is explained further in the section *"Key Aspects of AI Governance,"* where we concentrate on the conceptional and technical aspects of AI governance.

# Challenges for AI Governance

There are several distinct but still related domains or dimensions that play a central role regarding challenges for AI governance. These domains are naturally influencing and impacting each other.

The following list describes some sample domains[30], which we use to describe the challenges for AI governance:

- **Technology domain**: This includes AI systems and tools, DL and ML models and algorithms, AI applications, specialized AI engines and accelerators, AI solutions including robotics systems, and other AI technology.

---

[30]Additional domains are related to industry sectors, geographical differences, and human expertise in AI and governance, which – due to space constraints in this book – we won't elaborate on.

using clustering techniques, and entity extraction using ANNs are just a few examples of AI capabilities which could be applied in this step to determine whether or not a data source contains PII[24]. Leveraging such AI capabilities under the covers in this step, the data steward gets an assessment result whether or not the scanned source contains PII. The data steward can review the assessment result and decide whether or not the result was accurate by checking the prediction confidence or by reviewing relevant entries if necessary. In case the PII detection result was wrong, the data steward can correct the assessment.

4.  In the next step, data classes representing semantical concepts and business terms are assigned to the discovered attributes in a data source based on AI. The attribute metadata, the profiling results from the previous steps, and the data values are considered as input. For example, if a column based on semantical classification was detected as containing values representing cities, the data class representing the semantical concept of *city* would be assigned to that column, including the business term *city* describing with business metadata what a *city* is and how and why this attribute is used. Again, the data steward can review if the proposed assignments are accurate and approve or override them. For AI-based data class and business terms assignment, a data governance catalog needs to provide a different set of AI techniques to deal with the following cases effectively:

    a.  In case no labeled data is available, unsupervised learning techniques can be applied. This case applies when there are no data assets yet where attributes have assigned business terms.

    b.  Different application packages (e.g., SAP, Salesforce, etc.) or different custom applications might use very different naming conventions. So even if the AI-based term assignment looks at metadata describing the attributes as well as the actual

---

[24]Using machine learning presents risks to data privacy in its own right since trained models might be vulnerable to de-anonymization attacks possibly disclosing PII. You can learn more about this here [44].

attribute values, a model which works in one instance might
not work in another instance. In such a case, instance-based
learning methods with appropriate similarity measures need to
be applied for the AI-based term assignment.

c. A glossary of terms might have thousands or tens of thousands
of terms. So even with labeled data, this might turn into
a multi-class, multi-label classification problem very fast
with maybe only 5–50 instances per class. To overcome this
challenge, techniques like nearest neighbor or transforming
the problem into a binary classification problem can be used.

For steps 2–4 in this process, active learning capabilities can be applied to
continuously improve the results of the AI-based capabilities. The value proposition of
this AI-infused, automated process is very significant. For example, in step 4, without the
AI-based suggestions for data class and term assignments, a data steward would have a
need to search for appropriate data classes and terms in the catalog before being able to
establish the relationships between them and the attributes. Entering the keywords to
search for data classes or terms, reviewing the result lists, selecting the applicable entry,
and then creating the relationship to the attribute in question is time consuming. With
the AI capabilities in place, the data steward in vast majority of the cases is done with
a single mouse click approving the suggestion, taking out a lot of time and saving labor
costs. While there is a lot more to discuss on this topic, we hope with this introduction
you got a basic understanding why it is very valuable to apply AI techniques to metadata
and data quality management. Let's now move to another data governance area where
an infusion of AI capabilities was done in recent years which is data security in the next
section.

## AI Applied to Data Security

Data security is a critical aspect of data governance. The 2019 Cost of Data Breach
Report[25] indicates that the global average for a data breach in the USA is $3.9 million
with an average cost per case of $8.19 million. The cost is typically divided into direct
costs (fines for noncompliance with regulations, settlements, etc.) and indirect costs

---

[25]See [45] for the report, which contains a lot more details on the various dimensions of costs and
others.

(reputational damage, etc.). If data breach costs are expensive, the obvious next question is related to signs regarding a trend of increased data security issues. According to the IBM X-Force Threat Intelligence Index 2020,[26] the trend for 2020 and beyond shows an anticipated growth of security-related incidents. Thus, the problem of potential data breaches is not going away anytime soon. The largest source of data-related breaches (approximately 50%) are external attackers. Security professionals are urgently needed to help prevent incidents or – if an incident occurs – identify and remediate the incident as quickly as possible. But how easy is it that enterprises can hire security professionals for their IT staff? According to this ISACA report on the State of Cybersecurity 2020,[27] 57% of the enterprises in their survey indicated to have open positions for security positions where 30% are open for more than 3 months and 29% are open for more than 6 months until suitable candidates were found.

***Table 8-1.*** *Security Areas Infused with AI*

| Security Area | Key Value Points of AI |
| --- | --- |
| Data protection / security | Improved protection of data detecting and preventing illegal access to data |
| Security operation center | Prioritize alerts, classify indicents according to MITRE classification, collect relevant information and automatically recommend best action for the security analyst |
| Digital identity and fraud detection | With the help of AI, identify and prevent attacks on digital identies or fraud |
| Endpoint management | Leverage AI to protect your endpoints more effectively against attacks |
| User behavior | Leverage AI to detect anomalies in user behavior indicating potential threats and act on them |
| Intelligent incident response | Use AI for intelligent decision making in selecting and orchestrating the right incident responses |

---

[26]This report can be found here [46] and provides a lot of useful details on the various categories of attack, a breakdown in internal vs. external threats, and so on.

[27]The report containing the cited details can be found here [47].

This as background suggests three key questions. First, do attackers leverage AI for more sophisticated attacks? Yes, they do. For example, ANNs are introduced to improve password crackers. Second, are AI models or the data they were trained with the target of a potential attack? The answer is yes, as attackers steal AI models and potentially expose privacy by trying to steal training data which might contain PII data by using model inversion attacks. Third, can AI be used to protect you against security and data security threats in particular? Again, the answer is yes, and Table 8-1 shows you some relevant areas where AI-powered solutions are available today. The idea of Table 8-1 is to show you where in security, and in particular data security, novel AI-based solutions might be able to effectively[28] help you.

Software vendors[29] have offerings for the preceding areas available today. As part of your data governance program, particularly related to data security, you should thus look into what is needed to secure your data assets appropriately, which might require the use of several AI-infused software capabilities in the security area.

With a basic understanding of data governance, we can now look at the impact of how AI can be governed with the capabilities from the data governance space. We will also show which additional capabilities will be required in the data governance domain to extend it so that it can be effectively used in the AI domain as well.

# Governance in the Context of AI

Governance in the context of AI can be very pervasive and broad, reaching far beyond the pure technical aspects, which are just a means to get it right; governance in its roots are policies, standards, and compliance aspects that an organization should uphold.

In this section, we show the challenges for AI governance and how traditional information governance will be impacted by AI and machine learning. In addition, we briefly elaborate on some regulations driving AI governance for a trustworthy AI.

---

[28]AI is not solving everything in IT or data security in particular. There are many critical voices such as [48] indicating that you need to carefully assess the capabilities provided by software vendors to see where they provide reliable value adds and where you still need to rely on your security team expertise.

[29]For example, IBM offers IBM Guardium, IBM QRadar Advisor with Watson, IBM QRadar User Behavior Analytics, IBM Resilient, IBM Trusteer, and IBM MaaS360 with Watson to address all these areas with AI-infused capabilities.

# Beyond Traditional Information Governance

In this section, we elaborate on the mutual interference and impact of information governance and AI. Under the influence of AI, information governance needs to become broader by including new AI-related artifacts as well as AI-related challenges, extensions within the scope of governance. As we have seen in the previous sections, this causes challenges for application and enactment of AI governance. On the other hand, AI increases confidence in the data quality dimension of information governance by simply leveraging AI and machine learning algorithms and techniques, for instance, by increasing trustworthiness and accuracy for data matching tasks as you will see in Chapter 9, "*Applying AI to Data Governance and MDM.*"

AI represents a significant opportunity to advance some information governance aspects. For instance, AI accelerates insight and reduces the circle of action; after the corresponding action has been performed, it also reduces the time to realize the impact and revert or persist the change. AI furthermore helps in answering the questions fast instead of a human trying to read through a whole database.

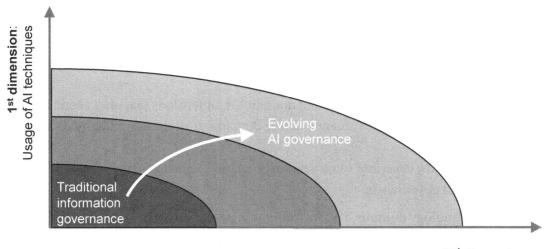

***Figure 8-6.*** *Interplay of Information Governance and AI*

Figure 8-6 illustrates the two-dimensional characteristic for traditional information governance to develop into AI governance:

- **First dimension**: Growth of information governance by leveraging AI and ML techniques. In this dimension, AI and ML constitute an improvement of information governance. As previously shown in this chapter, ML can significantly reduce the amount of human labor in source analysis and classification. Another example is to apply ML methods to identify operational risk patterns.

- **Second dimension**: Growth of information governance driven by the need to govern additional AI and ML artifacts, scenarios, terms and definitions, rules, and policies.

Without making the picture overly complicated, we could even add a third dimension to take into consideration evolving organizations climbing up their ladder to AI, offering new opportunities and new challenges, which need to be addressed and entertained by the AI governance framework. This evolving AI governance scope is explained further in the section *"Key Aspects of AI Governance,"* where we concentrate on the conceptional and technical aspects of AI governance.

# Challenges for AI Governance

There are several distinct but still related domains or dimensions that play a central role regarding challenges for AI governance. These domains are naturally influencing and impacting each other.

The following list describes some sample domains[30], which we use to describe the challenges for AI governance:

- **Technology domain**: This includes AI systems and tools, DL and ML models and algorithms, AI applications, specialized AI engines and accelerators, AI solutions including robotics systems, and other AI technology.

---

[30]Additional domains are related to industry sectors, geographical differences, and human expertise in AI and governance, which – due to space constraints in this book – we won't elaborate on.

- **Ethical and social domain**: This addresses ethical principles, guidelines, and generally accepted human and social values that AI needs to adhere to in order to gain greater confidence[31] and to remain human-centric, explainable, and fair to individuals and to increase acceptance by individuals and the society as a whole.

- **Political and legal domain**: Regulatory authorities and national governments[32] will create a body of laws and rules, policies, and regulations that are driven by technology and guided – possibly even limited – by ethical and social principles.

More than with any previous advances in technology, such as cloud computing or big data with its social media interaction, AI and machine learning have already begun to impact GRC. AI governance is aiming for a framework or model that addresses ethical and social, political and legal, and technical aspects in order to ensure the usage of AI in a human-centric and safe way.

Table 8-2 lists just a few of these challenges (without claiming completeness) and links them to the domains that we have discussed previously. These domains can be adjusted and further refined. The technology domain, for instance, can be segmented into robotics systems, application components, DL and ML models, and others. Depending on the specific AI scope that an organization is concerned about, additional domains have to be added, or existing domains have to be adjusted.

***Table 8-2.*** *Challenges for AI Governance*

| # | Challenge | Domain |
|---|---|---|
| 1 | Missing or insufficient transparency in AI algorithms and methods, the autonomous decision-making process, AI tools and applications | Technology |
| 2 | Incorrect decisions leading, for instance, to faulty life-threatening medical diagnosis and treatment, traffic accidents, and others | Legal |
| 3 | Missing standards and principles, guiding and limiting AI application scope and deployment | Political and legal |

*(continued)*

---

[31]See [49] for the *AI in control* framework from KPMG.

[32]We provide a few examples in the section "*Regulations Driving AI Governance*."

***Table 8-2.***  (*continued*)

| # | Challenge | Domain |
|---|-----------|--------|
| 4 | Incorrect decisions violating equal opportunity or adversely affecting gender or age of people | Ethical and social |
| 5 | Missing trust in AI systems, solutions, applications, and others – similar to missing trust in data | Political and social |
| 6 | Failures of AI robotics systems, accelerators, or AI applications in life-threatening situations | Technology and legal |
| 7 | Issues widening information governance to include AI-specific artifacts, for example, DL and ML models, whom to blame in case of autonomous car accidents, and others | Technology and legal |
| 8 | Learning aspects (e.g., of DL models) to be included within the scope of AI governance | Technology and ethical |
| 9 | Adjusting provenance (ownership) and data lineage to accommodate AI artifacts, for example, DL and ML models | Technology |
| 10 | Additional security risks that are imposed by self-learning of AI systems and applications | Technology and social |

For some of these challenges, the assignment to one or several of the domains is rather fuzzy, with some challenges even related to all domains.

# Regulations Driving AI Governance

In addition to the impact that AI has on organizations, regulations are one of the key driving factor for AI governance. In this section, we give a very short – although rather incomplete – overview on AI governance-related regulations. In most of the countries, there exist studies, exploratory efforts, and proposed regulatory statements. In this section, we provide only a rather brief overview on some initiatives, without asserting a claim for completeness.

In 2018, 25 European countries signed a declaration of cooperation on AI[33] which resulted in an EU regulation in February 2020 complementing national initiatives of some EU Member States. In addition to ensuring Europe's competitiveness in AI R&D matters, the goal is to address *social, economic, ethical, and legal questions.*

The German Government adopted an AI strategy[34] in 2018, which pursues *"safeguarding the responsible development and use of AI which serves the good of society, and integrating AI in society in ethical, legal, cultural and institutional terms in the context of a broad societal dialogue and active political measures."* In 2017, the German Federal Ministry of Transport and Digital Infrastructure published a report on automated and connected driving,[35] which includes ethical rules for automated and connected vehicular traffic. In 2017, the US Securities and Exchange Commission (SEC) published information and guidance[36] on robo-advisers for investors and the financial services industry on the fast-growing use of robo-advisers. In 2020, the Singapore Federal Government released an AI governance framework[37] *"to provide detailed and readily implementable guidance to private organizations on how to address ethical and governance issues when using AI solutions."*

## Key Aspects of AI Governance

In this section, we elaborate on the technical aspects of information governance that need to be adjusted and expanded to comprise AI aspects. In order to do this, we will take the core architecture building blocks (ABBs) of the information governance and information catalog layer of the AI information architecture from Chapter 4, *"AI Information Architecture,"* as a base. We will concentrate on the following subset of the ABBs: rules and policies, glossaries, search and discovery, classification, and provenance and lineage. The ABBs that are related to master data management and collaboration won't be described here.

---

[33]See [50] for more information on the European Commission Declaration of Cooperation on AI and the AI regulation which was released in February 2020 is located here [51].

[34]See [52] for more information on the Federal Government of Germany adopting an AI Strategy.

[35]See [53] for more information on the German Federal Ministry of Transport and Digital Infrastructure report on Automated and Connected Driving.

[36]See [54] for more information on the publication of the SEC on robo-advisers.

[37]See [55] for more information on the Model AI Governance Framework.

Figure 8-7 depicts the AI governance layer of the AI information architecture. For the complete AI information architecture including the ABBs from other layers, you should revisit Chapter 4, "*AI Information Architecture.*"

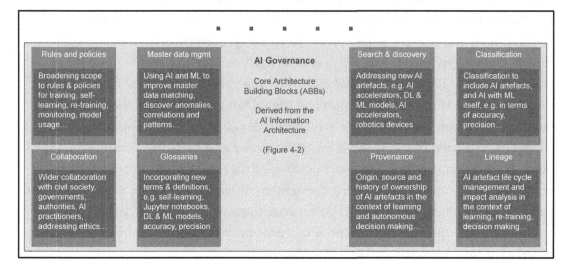

***Figure 8-7.*** *AI Governance*

In the following sections, the focus is on how information governance will change when you have to implement it in an AI-driven organization and AI-based world. However, we also briefly highlight the advantages using AI for information governance improvements.

# Rules and Policies

With all branches and cognitive aspects of AI, including supervised and non-supervised learning, reinforcement learning, and the corresponding application of DL and ML algorithms, new AI-based methods will be used and new activities will be performed. Some examples are training and regular retraining for ML models or the self-learning aspects of DL models that enable AI solutions, applications, and robotics devices to change and adjust in an autonomous fashion over time.

New AI artifacts such as DL and ML models, features, data assets, notebooks, and RESTful APIs have to be used and deployed within AI applications and enabled for usage in a regulated way. The monitoring of required accuracy and precision levels of ML models to ensure continuing business relevance constitutes a new set of tasks, which need to be governed as well.

Some of these methods and activities need to be governed and regulated via rules and policies, advocating the usage of an *AI guarding* agent. These rules and policies may be driven by pure technical aspects, for instance, ensuring continuing accuracy and precision of ML models. Business aspects, such as permission of AI artifact usage by various users, including management of roles and responsibilities, may have to be factored in. Ethical concerns and various regulations that may dictate usage patterns and limitations of usage scenarios may play a role as well. Rules and policies are largely impacted by AI and driven by technical, business, and the broader GRC accountability.

AI methods and techniques may also be leveraged in order to improve management, implementation, and governance of rules and policies. For instance, ML algorithms can be used to discover useful correlations of various rules and policies to simplify management aspects. Although traditional business rules management systems have a declining role in today's organizations, integrating them may still increase – at least for the time being – the efficiency and relevance of the rules and policies building blocks.

## Glossaries

AI comes with a set of new business and IT-related terms and definitions. Incorporating these new terms and definitions, for example, self-learning, Jupyter or Zeppelin Notebooks, DL and ML models, training and retraining, and AI-specific KPIs[38] to measure the business relevance of models, such as accuracy and precision, ROC or PR curve, and drift of accuracy – just to name a few – requires significant enhancements of existing business and IT glossary capabilities. Some of these terms may already be known and defined in the existing enterprise glossary; however, they may have to be adjusted and may be used in a somewhat different context. Whatever technology you use to manage terms used in your organization, a central catalog or dictionary is needed so terms mean the same thing across the organization, or if it means something different, then it is properly conjugated, annexed, and referenced.

Business metadata for terms and definitions, IT, and operational metadata (addressing all AI assets and artifacts) to specify AI technical assets, to enable understanding and usage of rules and policies, and to enable additional AI governance functions, such as provenance and lineage, will have to be adjusted as well.

---

[38]See [56] for an explanation of the most important AI, DL, and ML terms.

In addition, the alignment between the various roles and responsibilities in an AI-driven organization requires readjustments. For instance, some of the technical AI terms, such as ROC or PR curve or drift of accuracy, need to be understood, measured, monitored, and even adjusted by the business. An AI glossary[39] needs to facilitate these needs.

## Search and Discovery

Search and discovery, which are well-known capabilities within the traditional information governance scope, need to be adjusted for AI governance. This is not only driven by the need to address new AI artifacts, for example, AI accelerators, DL and ML models, or robotics devices. The required adjustment is also driven by new AI solution and DL and ML model development patterns. Data scientists need to search for and discover relevant and representative data assets for development, training, validation, and adjustments (retraining) of DL and ML models or artifacts. To a large degree, this is an exploratory task, where contextual search and discovery is complemented with visual exploration and subsequent data preparation tasks.

Another example is the need to search for and discover the most suitable DL or ML models among tens and even hundreds of alternative ones that can be developed, optimized, and trained using different hyperparameters, kernels, algorithms, and ML methods. This particular capability may be delivered either through AI development or AI governance tools. IBM provides the Auto AI graphical tool[40] in IBM Watson Studio to address exactly this task for predictive modeling problems.

## Classification

Information and data classification[41] are long-standing, well-known, and indispensable aspect of GRC. AI governance requires a new dimension of classification for AI assets and AI intellectual capital (e.g., DL and ML models, AI methods, AI solutions, etc.). Classification in the context of AI governance immediately suggests to apply ML algorithms to improve classification requirements for existing assets as well as new AI artifacts as seen earlier in this chapter. This, of course, requires a sufficiently large number of candidates to make ML techniques applicable for such a task. Depending

---

[39]We are using the terms glossary, dictionary, and catalog interchangeably.
[40]See [57] for more information on the Auto AI graphical tool in IBM Watson Studio.
[41]In this context, we refer to classification as categorization of data.

on the data, applying ML algorithms for classification of assets could potentially yield improved accuracy and business relevance as well as automated discovery of new meaningful clusters or categories to improve classification. There are numerous types of classification, for example, content based, context based, usage based, security based, location based, quality based, or sensitivity based.

The additional angle is to see how classification from the data governance space needs to be extended to classify and provide the right taxonomies for the AI artefacts which are new capabilities needed for AI governance. To adapt classification to embrace AI requires the coverage of the AI artifacts themselves, but also to apply additional types of classification that are, for instance, related to supervised vs. non-supervised models, type of regression models, and accuracy and precision of DL and ML models.

In this context, classification goes beyond a simple assignment of an AI artifact as an *ML model*; it means to take into account, for instance, the regression type (linear, stepwise, polynomial), training and validation method(s) used, pruning strategy used, methods used to reduce large sets of possibly correlated components or dimensions (predictors, features), accuracy or precision and their variations over the life cycle of the ML model, and so on.

## Provenance and Lineage

There is a strong affinity, even some overlap between provenance and lineage. In a way, these terms represent two sides of the same coin. Provenance is primarily concerned about the origin, source, and history of ownership of AI artifacts, especially in the context of learning, self-improving DL models, automated retraining of ML models, and autonomous decision making. An example is the ownership of an ML model, including the creating organization or individual who is allowed to use or change the model.

Lineage is more concerned about the life cycle management aspects of AI artifacts, including activities performed against those artifacts and impact analysis in the context of learning, retraining, and decision making. An example is the retraining of an ML model with new data, including why was retraining required or meaningful, what new labeled data sets were used for retraining, when was the model redeployed, and so on.

Another example is to understand biased outcome in terms of an ML model to show favorable or unfavorable outcome (bias) related to a specific feature or set of features[42]

---

[42]Please note that bias may not always be an unfavorable outcome; it may also be useful to generalize better in cases where larger data sets with various additional attributes are available.

and to analyze and visualize the impact of, for instance, adjusted feature weights to mitigate bias. Provenance and lineage have already been referenced as key ABBs in our AI information architecture in Chapter 4, "*AI Information Architecture*."

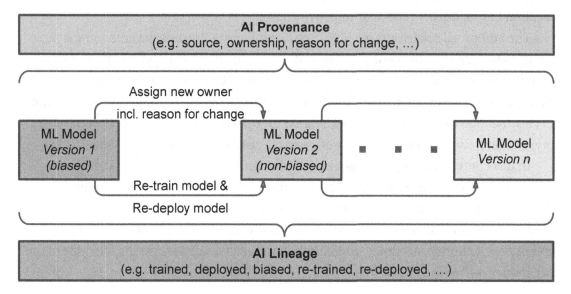

***Figure 8-8.*** *AI Provenance and Lineage*

Figure 8-8 is an exemplary illustration of the coexistence of AI provenance and lineage. It also illustrates the coexistence and individual role of these ABBs within the life cycle of an ML model.

# Mapping to Sample Vendor Offerings

There are quite a number of vendor offerings, products, and principles that can provide added value for AI governance. In this section, we introduce some key offerings and principles. However, instead of providing a comprehensive review, we concentrate on the relevance for AI-specific governance aspects. For the interested reader to gain a more in-depth understanding, we provide links to respective documentation.

As you can imagine, the following list is far from complete:

1. *Amazon: Amazon Web Services with Amazon SageMaker*

2. *Microsoft*: Microsoft AI principles, Microsoft Bot Framework, and Azure Cognitive Services

3. *IBM offerings*: Watson Knowledge Catalog, Watson OpenScale, and OpenPages with Watson

# Amazon Web Services (AWS)

In recent years, Amazon has made significant enhancements to enrich their Amazon Web Services (AWS) with AI and ML capabilities. This includes AI governance enhancements, which are primarily delivered via the Amazon SageMaker[43] platform. Amazon SageMaker is a fully managed ML service, which can be used by data scientists and developers to build and train ML models and then directly deploy them into a production-ready hosted environment for predictive or analytical applications in the AWS public cloud.

Amazon SageMaker comes with a rich set of features. The following list includes those features, which are related – in a wider sense – to ML life cycle management and AI governance needs:

- **Amazon SageMaker Model Monitor**: To monitor and analyze ML models in production (endpoints), to detect data drift and deviations in model quality. This is a service that detects concept drift in models and alerts developers when the performance and accuracy of a deployed ML model begin to deviate from the original trained model.

- **Amazon SageMaker Debugger**: Comes as part of the SageMaker Studio to inspect training parameters and data throughout the training process. Automatically detects and alerts users to commonly occurring errors such as parameter values getting too large or small.

Amazon has made first steps toward ML model explainability through the visualization of the metrics from SageMaker Debugger in the SageMaker Studio for further interpretation. SageMaker Debugger can also generate warnings and remediation advice that may occur within the ML model training phase. With SageMaker Debugger, you can interpret how a model is working, representing an early step toward ML model explainability.

---

[43]See [58] for more information on Amazon SageMaker.

# Microsoft AI Principles

Microsoft is at the forefront in integrating AI into some of their offerings and services. This is, for instance, clearly visible in Microsoft's offerings, such as Skype, Cortana, Bing, or Office 365. This relates to embedded chatbots, translations services, and other AI services that become available to millions of end users. Microsoft is not only leveraging AI technologies to enhance their own product portfolio; they are also making AI technologies available to developers for them to develop their own AI-infused offerings.

AI governance is a vital component of infusing AI services into offerings. This becomes obvious by taking a closer look at the following Microsoft AI principles[44]:

1. **Fairness**: AI systems should treat all people fairly.

2. **Reliability and safety**: AI systems should perform reliably and safely.

3. **Privacy and security**: AI systems should be secure and respect privacy.

4. **Inclusiveness**: AI systems should empower everyone and engage people.

5. **Transparency**: AI systems should be understandable.

6. **Accountability**: AI systems should have algorithmic accountability.

These Microsoft AI principles can be mapped to the key aspects of AI governance, which we have described earlier.

Complementing guidelines for developing responsible conversational AI, Microsoft offers tools, such as the Microsoft Bot Framework[45], which is a comprehensive framework for building enterprise-grade conversational AI experiences. The Azure Bot Service does not only enable you to build intelligent, enterprise-grade bots but also to integrate this with other Azure Cognitive Services.

---

[44]See [59] and [60] for more information on the Microsoft AI principles.
[45]See [61] for more information on the Microsoft Bot Framework.

# IBM Offerings

IBM Watson Knowledge Catalog[46] infuses governance and enables self-service for finding, accessing, and preparing data in essence enabling them for AI with DL and ML. It consists of policy management to enable access, curation, categorization, and sharing of data, knowledge assets, and their relationships. With information governance, data quality, and active policy management, it helps your organization protect and govern sensitive data, trace data lineage, and manage AI at scale.

The Watson Knowledge Catalog can be integrated or automatically synchronized with the IBM InfoSphere Information Governance Catalog to simply use existing assets of the Information Governance Catalog. Users can create a comprehensive business and IT glossary that helps to bridge the communication gap between LoB and IT organizations. Glossary assets are terms, categories, and governance policies and rules. In addition to the glossary, you can also manage metadata about information assets. The Watson Knowledge Catalog is part of the larger Unified Governance and Integration (UGI) portfolio from IBM[47].

IBM Watson OpenScale[48] allows you to govern and operationalize AI deployments, while tracking and measuring outcome from AI across its life cycle. It provides insight around the performance and accuracy of AI models and ensures transparent, explainable, and fair outcome that is freed from bias. Watson OpenScale bridges the gap between data scientist and application development teams, who have to provide business units with AI-infused applications, and IT teams, who have to deploy and operationalize models and scoring applications.

The following is a list of some key features. These capabilities address some of the most important AI governance aspects identified in this chapter:

- **Generating trusted and explainable AI outcome** in understandable business terms, so that business users can explain how an AI model arrived at a prediction. Watson OpenScale includes feature weighting (feature importance shown in terms of weights) to understand the most and least important features and explains in terms of the top K features which played a key role in the prediction.

---

[46]See [38] for more information on the IBM Watson Knowledge Catalog.
[47]See [62] for more information on IBM Unified Governance and Integration.
[48]See [63] for more information on IBM Watson OpenScale.

- **Ensuring AI model fairness** by continuously monitoring how AI makes predictions. You can select the feature(s) of an AI model to monitor the deployed model's propensity for a favorable or unfavorable (biased) outcome. The output could, for instance, indicate that the model is biased on a specific value or range of a feature (e.g., gender=female, age>45).

- **Enabling traceability and auditability** of AI predictions via scoring modules or applications for the entire life cycle of AI assets, starting from the initial design, training, validation, and deployment through to operation, monitoring, retraining, and end of life. For instance, for a given ML model, it allows to identify teams and ownership to be assigned, data used for training, and outcome produced.

- **Measuring model accuracy** to understand the quality and business relevance of the model during production. Using labeled data, Watson OpenScale measures – among other things – the area under the ROC or PR curve but also detects the drift without labeled data of the drop in accuracy and in data consistency.

IBM OpenPages with Watson[49] provides a cognitive-driven GRC portfolio that is comprised of various offerings. It provides a set of core services and functional components addressing operational risk, regulatory compliance, IT risk and security, internal audit, vendor risk management, and financial controls management.

# Key Takeaways

We conclude this chapter with a few key takeaways, summarized in Table 8-3. As we have seen, AI does not only have a profound impact on governance; there is a mutual interference between information governance and AI: there is a need to govern new AI artifacts, but AI governance itself will see a benefit from AI. The key aspects of AI governance as pointed out in this chapter need to be tailored to your enterprise depending on your particular focus and requirements.

---

[49]See [64] for more information on IBM OpenPages with Watson.

***Table 8-3.*** *Key Takeaways*

| # | Key Takeaway | High-Level Description |
|---|---|---|
| 1 | There is a need to govern AI artifacts | AI artifacts such as DL and ML models, AI solutions, robotics devices, and others drive for an adjusted scope to AI governance |
| 2 | AI governance itself will benefit from AI | Leveraging AI and ML algorithms and methods, for example, by increasing trustworthiness and accuracy for data matching tasks |
| 3 | New regulations drive AI governance | These are local and global efforts and recommendations that are largely driven by ethical, cultural, and legal concerns |
| 4 | There are a number of vital AI challenges | These challenges can be assigned to one or several of the following domains: technology, ethical and social, or political and legal |
| 5 | Metadata management as part of data governance catalog tooling must be AI infused to be efficient | Using AI to create new business terms from a fast-growing set of regulations and unstructured sources accelerates glossary authoring and reduces manual labor for data stewards. AI-infused term assignment for curating new data sources as governed data assets is another critical accelerator to keep the workload for data stewards manageable |
| 6 | Data quality management as part of data governance tooling must be AI infused | Using AI to learn new data rules automates the discovery of new data quality rules. Using AI to suggest fixes of data quality issues accelerates or even automates the resolution of some data quality issues reducing the labor time to do so manually |
| 7 | Using AI to discover PII in structured and unstructured data is a critical step for your journey toward compliance with many privacy regulations | Numerous privacy regulations mandate the appropriate management of PII. Noncompliance with GDPR, for example, poses a significant financial risk. As a first step, PII data must be identified, which is particularly challenging for unstructured sources. With more data being on public cloud, the brute force-discovery using APIs for indexing is impractical. The AI-based discovery of PII must hence not only be able to deal with a broad variety of different data formats and systems for which connectors are required but must also be combined with efficient sampling methods |

(*continued*)

***Table 8-3.*** *(continued)*

| # | Key Takeaway | High-Level Description |
|---|---|---|
| 8 | AI in data security is a critical capability for prioritizing and responding to the right security incidents | In the security operations center (SOC), security professionals are overwhelmed with the fine-grained but very large volumes of security-related information flowing in. AI helps to focus the attention of the security professionals to the real and important threats for the security of data so that they can respond to the events which are critical keeping your data safe |

# References

[1]   Weinstein, S. *Legal Risk Management, Governance and Compliance.* ISBN-13: 978-1909416512, Global Law and Business Ltd., 2016.

[2]   General Data Protection Regulation *(GDPR):* https://eur-lex.europa.eu/eli/reg/2016/679/oj *(accessed May 2020).*

[3]   Health Insurance Portability and Accountability Act (HIPAA). www.hhs.gov/hipaa/index.html (accessed August 23, 2019).

[4]   *Sarbanes-Oxley Act* (SOX). www.sec.gov/news/studies/2009/sox-404_study.htm (accessed August 23, 2019).

[5]   Sen, Harkish. *Data Governance: Perspectives and Practices.* ISBN-13: 978-1634624787, Technics Publications, 2019.

[6]   General Data Protection Regulation *(GDPR):* https://eur-lex.europa.eu/eli/reg/2016/679/oj *(accessed May 2020).*

[7]   *California Consumer Privacy Act (CCPA):* https://oag.ca.gov/privacy/ccpa *(accessed May 2020).*

[8]   *Basel Committee on Banking Supervision Standard 239 (BCBS 239): Principles for effective risk data aggregation and risk reporting.* www.bis.org/publ/bcbs239.htm *(accessed May 2020).*

[9]   *Solvency II:* https://eur-lex.europa.eu/legal-content/EN/ALL/?uri=celex%3A32009L0138 *(accessed May 2020).*

[10]   *MiFID II:* `https://eur-lex.europa.eu/legal-content/EN/ALL/?`
       `uri=celex%3A32014L0065` *(accessed May 2020).*

[11]   *Protection of Personal Information Act (POPI):* `www.justice.gov.`
       `za/inforeg/docs/InfoRegSA-POPIA-act2013-004.pdf` *(accessed*
       *May 2020).*

[12]   *John Ladley: Data Governance: How to Design, Deploy and Sustain*
       *an Effective Data Governance Program. ISBN-13: 978-0124158290,*
       *Morgan Kaufmann, 2012.*

[13]   *David Plotkin: Data Stewardship. An Actionable Guide to*
       *Effective Data Management and Data Governance.*
       *ISBN-13: 978-0124103894, Morgan Kaufmann, 2013.*

[14]   *Harkish Sen: Data Governance: Perspectives and Practices.*
       *ISBN-10: 1634624785, Technics Publications, 2018.*

[15]   *Sunil Soares: The Chief Data Officer Handbook for Data*
       *Governance. ISBN-13: 978-1583474174, MC Press, 2015.*

[16]   *Sunil Soares: Big Data Governance: An Emerging Imperative.*
       *ISBN-13: 978-1583473771, MC PR LLC, 2012.*

[17]   *Neera Bhansali: Data Governance. Creating Value from Information*
       *Assets. ISBN-13: 978-1439879139, Auerbach Publications, 2013.*

[18]   *Rupa Mahanti: Data Quality: Dimensions, Measurement, Strategy,*
       *Management and Governance. ASIN: B07QMNT6HM, Amazon Media.*

[19]   *DAMA International:* `https://dama.org/content/what-data-`
       `governance` *(accessed May 2020).*

[20]   *The Data Governance Institute:* `www.datagovernance.com/adg_`
       `data_governance_definition/` *(accessed May 2020).*

[21]   *Commonwealth Enterprise Information Architecture:* `www.vita.`
       `virginia.gov/media/vitavirginiagov/it-governance/ea/pdf/`
       `Commonwealth_EIA_Strategy_FINAL.pdf` *(accessed May 2020).*

[22]   *Mario Godinez, Eberhard Hechler, Klaus Koenig, Steve Lockwood,*
       *Martin Oberhofer, Michael Schroeck: The Art Of Enterprise*
       *Information Architecture. A Systems-based Approach for Unlocking*
       *Business Insight. ISBN-13: 978-0137035717, IBM Press, 2010.*

[23]    *The Open Group Architecture Framework (TOGAF):*
        `www.opengroup.org/togaf` *(accessed May 2020).*

[24]    ISO/IEC 27000: `www.iso.org/standard/73906.html` *(accessed May 2020).*

[25]    ISO/IEC 27001: `www.iso.org/standard/54534.html` *(accessed May 2020).*

[26]    ISO/IEC 27002: `www.iso.org/standard/54533.html` *(accessed May 2020).*

[27]    *Software AG - Alfabet:* `www.softwareag.com/be/products/aris_alfabet/eam/default.html` *(accessed May 2020).*

[28]    *LeanIX Enterprise Architecture Suite:* `www.leanix.net/en/` *(accessed May 2020).*

[29]    *Spark Systems – Enterprise Architect:* `https://sparxsystems.com` *(accessed May 2020).*

[30]    *ArchiMate:* `www.archimatetool.com` *(accessed May 2020).*

[31]    *Lowell Fryman, W.H. Inmon, Bonnie O'Neill: Business Metadata: Capturing Enterprise Knowledge. ASIN: B003VWBXYG, Morgan Kaufmann, 2010.*

[32]    *Jian Qin, Marcia Lei Zeng: Metadata. ISBN-13: 978-1783300525, Facet Publishing, 2016.*

[33]    *Geradus Blokdyk: Reference Data Management. A Clear and Concise Reference. ISBN-13: 978-0655320333, 5STARCooks, 2018.*

[34]    *John Baldwin, Whei-Jen Chen, Thomas Dunn, Mike Grasselt, Shabbar Hussain, Dan Mandelstein, Erik O'Neill, Sushain Pandit, Ralph Tamlyn, Fenglian Xu: A Practical Guide to Managing Reference Data Management with IBM InfoSphere Master Data Management Reference Data Management Hub. IBM Redbooks, O'Reilly, 2013.* `www.oreilly.com/library/view/a-practical-guide/0738438022/` *(accessed May 2020).*

[35]   *Malcolm Chisholm: Managing Reference Data in Enterprise Databases. ISBN-13: 978-1558606975, Morgan Kaufmann, 2000.*

[36]   *Ivan Milman, Martin Oberhofer, Sushain Pandit, Yinle Zhou: Principled Reference Data Management for Big Data and Business Intelligence. In: International Journal of Organizational Collective Intelligence. 2017.* `https://doi.org/10.4018/IJOCI.2017010104`.

[37]   *Collibra – Data Catalog:* `www.collibra.com/data-catalog` *(accessed May 2020).*

[38]   *IBM – Watson Knowledge Catalog:* `www.ibm.com/cloud/watson-knowledge-catalog` *(accessed May 2020).*

[39]   *Data Interoperability Standards Consortium:* `https://datainteroperability.org` *(accessed May 2020).*

[40]   *Amir Efrati, Kevin McLaughlin: AWS Customers Rack Up Hefty Bills for Moving Data.* `www.theinformation.com/articles/aws-customers-rack-up-hefty-bills-for-moving-data` *(accessed May 2020).*

[41]   *IBM Guardium:* `www.ibm.com/products/ibm-guardium-data-protection/resources` *(accessed May 2020).*

[42]   *Alexander Borek, Martin Oberhofer, Ajith Kumar Parlikad, Philip Mark Woodall: A classification of data quality assessment and improvement methods. In: ICIQ 2011 – Proceedings of the 16ᵗʰ International Conference on Information Quality, p. 189-203, 2011.*

[43]   *CDQ Good Practice Award 2018 for Bosch:* `www.cc-cdq.ch/sites/default/files/cdq_award/CDQ%20Good%20Practice%20Award%202018_Robert%20Bosch.pdf` *(accessed May 2020).*

[44]   *Arianna Dorschel: Data Privacy in Machine Learning.* `https://luminovo.ai/blog/data-privacy-in-machine-learning` *(accessed May 2020).*

[45]    *2019 Cost of Data Breach Report:* https:// databreachcalculator.mybluemix.net/ *(accessed May 2020).*

[46]    *IBM X-Force Threat Intelligence Index 2020:* www.ibm.com/security/ data-breach/threat-intelligence *(accessed May 2020).*

[47]    *ISACA – State of Cybersecurity 2020:* www.isaca.org/bookstore/ bookstore-wht_papers-digital/whpsc201 *(accessed May 2020).*

[48]    *Lily Hay Newman: AI Can Help Cybersecurity – If It Can Fight Through the Hype.* www.wired.com/story/ai-machine-learning-cybersecurity/ *(accessed May 2020).*

[49]    KPMG. *KPMG Artificial Intelligence in Control – Establish greater confidence in your AI technology performance.* https://home.kpmg/ xx/en/home/insights/2018/12/kpmg-artificial-intelligence-in-control.html (accessed September 6, 2019).

[50]    European Commission. *EU Member States sign up to cooperate on Artificial Intelligence.* https://ec.europa.eu/digital-single-market/en/news/eu-member-states-sign-cooperate-artificial-intelligence (accessed August 28, 2019).

[51]    European Commission: *On Artificial Intelligence – A European approach to excellence and trust.* https://ec.europa.eu/info/ sites/info/files/commission-white-paper-artificial-intelligence-feb2020_en.pdf (accessed April 2020).

[52]    German Federal Ministry for Economic Affairs and Energy. *Federal Government adopts Artificial Intelligence Strategy.* www. bmwi.de/Redaktion/EN/Pressemitteilungen/2018/20181116-federal-government-adopts-artificial-intelligence-strategy.html (accessed August 28, 2019).

[53]    *German Federal Ministry of Transport and Digital Infrastructure – Ethics Commission. Automated and Connected Driving.* www. bmvi.de/SharedDocs/EN/publications/report-ethics-commission. pdf?__blob=publicationFile (accessed August 28, 2019).

[54]    U.S. Securities and Exchange Commission (SEC). Division of Investment Management. Guidance Update – Robo-Advisers. www.sec.gov/investment/im-guidance-2017-02.pdf (accessed August 28, 2019).

[55]    Personal Data Protection Commission Singapore. *Model AI Governance Framework*. www.pdpc.gov.sg/Help-and-Resources/2020/01/Model-AI-Governance-Framework (accessed July 8, 2020).

[56]    James, G., Witten, D., Hastie, T., Tibshirani, R. *An Introduction to Statistical Learning: with Applications in R (Springer Texts in Statistics)*. ISBN-13: 978-1461471370, Springer, 2013.

[57]    IBM. *AutoAI Overview*. https://dataplatform.cloud.ibm.com/docs/content/wsj/analyze-data/autoai-overview.html(accessed September 6, 2019).

[58]    Amazon. *Amazon SageMaker Developer Guide*. https://docs.aws.amazon.com/sagemaker/latest/dg/whatis.html (accessed December 5, 2019).

[59]    Microsoft. *Microsoft AI Principles*. www.microsoft.com/en-us/ai/our-approach-to-ai (accessed December 1, 2019).

[60]    Microsoft. *The Future Computed. Artificial Intelligence and its role in society*. https://3er1viui9wo3Opkxh1v2nh4w-wpengine.netdna-ssl.com/wp-content/uploads/2018/02/The-Future-Computed_2.8.18.pdf (accessed December 3, 2019).

[61]    Microsoft. *Microsoft Bot Framework*. https://dev.botframework.com/ (accessed December 3, 2019).

[62]    IBM. *IBM Unified governance and integration*. www.ibm.com/ae-en/analytics/unified-governance-integration (accessed August 31, 2019).

[63]    IBM. *IBM Watson OpenScale*. www.ibm.com/cloud/watson-openscale/ (accessed August 31, 2019).

[64]    IBM. *IBM OpenPages with Watson*. www.ibm.com/downloads/cas/QQMA8FOW (accessed August 31, 2019).

# CHAPTER 9

# Applying AI to Master Data Management

We introduced in Chapter 8, "*AI and Governance*," data governance and the use of AI capabilities making data governance smarter. A related capability used by many enterprises is master data management (MDM). Depending on the industry, vertical customer, person, organization, product, supplier, patient, employee, citizen, and asset are typical examples of master data entities. MDM is used to deliver a trusted 360° view of master data which helps many critical operational processes such as customer service, cross- and up-sell, consistent customer experience in a multichannel architecture, or streamlined new product introduction.

The 360° view from MDM is critical for many analytical applications as well. It is essentially the trusted dimensional data for reporting in an enterprise DWH or a required source for data science projects focused on customer insights in areas such as next best offer, next best action, customer segmentation, churn analytics or topic, and sentiment detection.

If clients decide to adopt MDM, it is not a one-off project. A decision for MDM means establishing a program driving continuous business improvements by rolling out MDM to the enterprise. A critical success factor for MDM programs is an effective data governance program. To make a 360° view trustworthy, it must have proper data quality, data lineage, and access controls in place which are all also aspects of data governance. Many master data types are also regulated, for example, patient data falls under the HIPAA regulation, and person data is affected by many privacy regulations such as GDPR, CCPA, and so on. In a nutshell, there are many compelling reasons why MDM and data governance programs intersect.

© Eberhard Hechler, Martin Oberhofer, Thomas Schaeck 2020
E. Hechler et al., *Deploying AI in the Enterprise*, https://doi.org/10.1007/978-1-4842-6206-1_9

In this chapter, we introduce key aspects of MDM followed by a section showing how the AI-infused data governance capabilities are used for MDM solutions today and which areas of MDM itself have been infused by AI. As a third major aspect of this chapter, we explain how MDM can be leveraged to operationalize AI-based customer insights effectively introducing the emerging concept of the digital twin.

In the following section, we briefly[1] introduce the key concepts of master data management (MDM).

# Introduction to Master Data Management

MDM is a discipline known to data management professional since around 2000, and the MDM market for software and services is estimated to grow to $7 to $8 billion total in the next 2–3 years depending on analyst firms. Many software vendors like Informatica, Tibco, IBM, SAP, Riversand, Semarchy, and many more provide mature software solutions for MDM. More and more MDM offerings become available as containerized solutions on Kubernetes platforms for private cloud consumption or as software as a service (SaaS) solutions on public cloud. Given the maturity of the MDM market, Figure 9-1 provides you a summary of the most common capabilities of MDM software solutions today.

| Configuration & admin UX | Entity maintenance & hierarchy UX | Product catalog management UX | Data stewardship UX |
| Configuration & admin API | CRUD & search API | Link / unlink API | Import / export API & notifications |
| Matching engine | Machine Learning (ML) | Task & workflow management | Job management |
| Graph engine | Persistency | Text search | Security |

**Master Data Management (MDM) Capabilities**

***Figure 9-1.*** *Master Data Management Capabilities*

The user experience (UX) of an MDM software solution typically has functions to configure and administrate the MDM software. The configuration functionality usually allows to create and customize the MDM data model being used. The administration options allow, for example, to set access privileges for users or turn on – if required – features for writing an audit trail of all data changes and service requests. The entity maintenance and

---

[1]There are a number of books [1], [2], [3], and [4] we recommend in case you want to study MDM.

hierarchy UX functionality allows to search, read, create, update, and delete master data records and entities (entities are also known as golden records; they are the result of matching being applied to the master data records).

Maintaining organization hierarchies, sales territory hierarchies, credit risk hierarchies, or product category hierarchies are just a few examples for which hierarchy management functions in the UX are used. Authoring new products and maintaining product catalogs is another functional area for which specialized UX components are required. Data stewards managing data quality for the MDM system require comprehensive task management and data quality remediation processes accessible in the data stewardship UX.

The UX functionality consumes a rich API layer for all functions required. However, the real-time CRUD (short for create, read, update, delete) and SEARCH APIs consumed by channel applications, such as online and mobile channels, CRM systems, call centers, and others carry the burden of the workload, with 85–90% of the workload from these systems being read requests and 10–15% being write requests. Notification management notifies subscribed applications if a write requests modified master data. The import and export APIs for bulk processing are used for initial bulk load of master data into the MDM system, delta batch processing (e.g., monthly third-party feeds from Dun & Bradstreet), or batch exports into enterprise DWH or DS environments.

The matching engine is used to detect whether or not records are duplicates; it provides such functions in bulk, incremental batch, and real-time mode. We explore the matching problem in more detail in the next section. When the matching engine determines whether or not two or more records are duplicates, the outcome can be a non-match, auto-match, or clerical decision, which means that some similarity is found but not enough for an auto-match resolution. In such a case, a task is created for the data steward where the need for task and workflow management arises. The task and workflow component in support of data quality management provides all aspects required for human workflows remediating the identified data quality issues. The ML component is the most recent component added to an MDM system. We explore how the ML component is used to infuse MDM with AI in the next section. There are tasks like initial load, bulk cross matching, or large exports which require job management functionality to schedule these long-running tasks; see their status and results.

Graph-based exploration of master data relationships based on a graph database engine, a persistency to store the data, and a text search backend supporting easy-to-use text search functionality are also common components in MDM systems.

Given that master data is considered to be among the most valuable data assets for any enterprise, and many master data domains such as person or patient are affected by regulations, a broad range of security features are also required. Typical security features required include encryption of master data at rest, support for encrypted communication (e.g., SSL) and optionally message encryption, fine-grained data access controls on record and attribute level, and full audit functionality of all data changes and read service requests.

# Digital Twin and Customer Data Platform

An emerging trend next to MDM is the rise of the concepts of the digital twin and customer data platform (CDP), which is a new software segment next to MDM that is in the Gartner Hype Cycle for technologies currently at the peak of inflated expectations. Compared to MDM, it is a small software market with an anticipated rapid growth. The driving motivation for CDP solutions is the concept of the digital twin, which is well understood in the IoT domain where the concept of the digital twin is used to create a digital representation of the physical fabric of the real world. Examples include assembly lines or utility networks (gas, electricity, water, etc.), which are fully instrumented using sensors. Applying the concept of the digital twin to customer master data is basically expanding the 360° view of the customer with many more attributes to get an enriched perspective on how the customer prefers to buy, get support, or to collect information about products or how the customer interacts on social platforms favorably around a product bought (or speaking negatively about it, if not happy with it).

It requires to link all customer interactions across all systems of engagements to the customer profile, where the customer can interact with an enterprise directly or indirectly. Figure 9-2 shows a conceptual view of the traditional MDM data model scope and extensions to it as required by the digital twin concept.

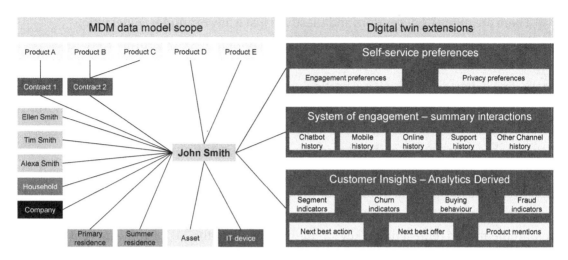

***Figure 9-2.*** *Extending the MDM 360° Customer View with Digital Twin Attributes*

Please note that this view is not a "black and white" one. For example, a couple of engagement preferences or privacy preferences have certainly been managed in MDM systems before. The key point here is that the depth of these preferences significantly increased. The comprehensive linking to all interaction touchpoints where the client engaged in via a system of engagement is another area which became increasingly important today. That does not mean you would push the voice recording from the call center into a MDM system, but you need to know that the customer called, the reason for the call, the outcome, who from your staff talked to the customer, and so on. This summary interaction information has to be linked to the customer profile to make it a digital twin profile.

Many companies invested already in AI to drive customer insights. However, without a strategy to operationalize these customer insights at the touchpoints where these insights could drive value, they are not of much use. In the last section of this chapter, we show how this problem can be solved.

Before doing so, let us compare strengths and weaknesses of CDP and MDM platforms as shown in Figure 9-3.

|  | Customer data platform<br>Buyer: leader in marketing | MDM<br>Buyer: leader in enterprise IT |
|---|---|---|
| Strength | ■ Connectivity<br>■ Segmentation<br>■ Activation | ■ Data Governance<br>■ Matching & deduplication<br>■ Stewardship |
| Weakness | ■ Data Governance<br>■ Matching & deduplication<br>■ Stewardship | ■ Connectivity<br>■ Segmentation<br>■ Activation |

***Figure 9-3.*** *CDP and MDM Comparison*

There are many CDP vendors and these are just a few examples: Blueconic, Segment, CrossEngage, QuanticMind, Scal-e, CustomerLabs CDP, Sprinklr Platform, InsideView, and Exponea. The CDP vendors are typically selling to key leaders in the marketing department, which are particularly interested in their strengths in customer segmentation and activation. Also, being able to see all the interactions with clients from all systems of engagements through a huge number of out-of-the-box connectors (e.g., Segment has over 300 out of the box) is appealing to marketing people.

The buyer of the MDM solutions is oftentimes in an enterprise IT department and must solve the master data problem for the enterprise as a whole. Hence, excellent data governance, matching and de-duplication, and stewardship capabilities are essential requirements and represent the key strengths of many of the MDM software vendors.

To avoid disruption, MDM vendors started to improve their MDM software solutions to better serve the digital twin requirements by increasing the flexibility to add relevant digital twin attributes with less effort to their MDM solutions. They now also allow to include summary interaction records into MDM systems. Another angle is improving the integration with the AI ecosystem, which we discuss in the last section of this chapter.

# Infusing AI into Master Data Management

The heart of MDM solutions is their capabilities to match and de-duplicate master data as well as their data stewardship capabilities. Known today are deterministic and fuzzy probabilistic matching techniques. Deterministic techniques are simple things like exact string comparison of attributes. However, master data was oftentimes created before an MDM system got deployed. The applications which created these records were not optimized to do so with proper data quality controls in place.

As a result, source data usually comes to MDM with different levels of data quality and a lot of data quality problems in it. Figure 9-4 shows you just the tip of the iceberg of data quality issues in two example records. In the fourth column, we listed some of the required fuzzy similarity techniques to assess if the two records shown are a match. In the last column, we show the weights of the attributes. The weights can be assigned to individual attributes (e.g., DOB, gender) or to a group of attributes (e.g., name attributes or address attributes). Not all attributes are equally important for the decision if two or more records are the same.

| Attribute | Record 1 | Record 2 | Similarity techniques | Weight |
|-----------|----------|----------|----------------------|--------|
| First name | Robert | Bob | Nickname resolution | 3.4 |
| Middle name | John | | | |
| Last name | Washington | Washincton | Phonetic similarity | |
| Suffix | Jr. | Junior | Standardization | |
| DOB | 10/03/1975 | 03/10/1975 | Date similarity (US vs. Europe) | 4.1 |
| Gender | Male | M | Standardization | 0.2 |
| SSN | 231-44-5821 | 231-45-5812 | Edit distance | -5.2 |
| Street | Sunset Boulevard | Sunset Blvd. 233 | Standardization for Blvd. and misplacement for 233 | 2.5 |
| House number | 234 | | | |
| ZIP | 91042 | 91042 | | |
| City | Los Angeles | Los Angeles | | |
| Country | USA | United States of America | Standardization | |

**Figure 9-4.** *Matching Techniques Required to Match Fuzzy Source Data*

For example, the gender column has usually a low number of distinct values. Finding a match of the values in this attribute is hence of low significance, and the weight contribution to the overall assessment of the compared records is small. Other attributes like the SSN are very important. In the example, the edit distance for the SSN is equal to 2 (two typos if you compare the values). When configuring the weighting schema for an attribute, you can assign negative or positive values. For example, to indicate when the fuzzy similarity is considered too far apart, this attribute should actually have a negative value, which means it makes a stronger contribution to a potential decision to keep the records apart. In the example shown, edit distance of 2 or more is configured with negative weights, edit distance of 0 would provide a large positive weight, and edit distance of 1 would still have some positive weight.

While based on various fuzzy comparison techniques, some similarity in the name and address attributes are found, they contribute less positive weight compared to values which would be more similar. Another aspect affecting the weights is the probability of a match. Consider a German data set. The last name *Cheng* compared to the last name *Müller* is very rare in a German data set, where for *Müller* you likely have a lot of entries. This means if you compare records on the last name, the probability to compare records with *Müller* in the last name is much higher compared to matching two records where the last name value is *Cheng*. Hence, the weights get a scaling factor applied. For rare values, the weights are scaled up, and for common values, the weights get scaled down since it is much more likely to find record with similarities for frequently occurring values. The total score in this example is 3, 4 + 4, 1 + 0, 2 + 2, 5 − 5, 2 = 5. The total score is usually compared to two thresholds. If the total score is below the lower threshold, it is a non-match. If the total score is above the upper threshold, it is an auto-match, and configurable survivorship rules would automatically create an entity record to which the duplicates are linked. If the total score is in between the lower and the upper threshold, this indicates that the records show some similarity, but a data steward needs to make the final decision whether or not the records are linked. In such a case, a task for the data steward is created.

Another aspect of matching is doing it at scale and with performance. Today, scale means billions of records in the largest customer implementations. This creates multiple challenges for matching. First, how can you run bulk matching at scale on very large data sets which is a requirement for the initial load of the MDM system? This problem is addressed with data partitioning and parallel execution. Second, if new records are added to MDM, or existing records are updated where you need to match again, how can you make that work with real-time performance measured in milliseconds? Obviously, brute-force comparison of a new record against every single record in a master data set of two billion records will not work in millisecond range. In addition, only a small subset of all the records have a chance to actually match. To find this small subset of potential candidates, where executing matching makes sense, is hence an interesting problem. To solve it, bucketing techniques are used, which in essence are techniques clustering the data where records with some commonality (e.g., same last name, same zip code) are located close to each other.

The candidate list is determined by comparing the relevant attributes of that record which must be matched against the cluster patterns, and the candidate list would only contain records where the values lined up. For example, if the record to match has the last name *Müller* and the zip code would be *70579*, only the records in the cluster given by *Müller, 70759,* would appear in the candidate list for matching. In practice, usually different attributes groups would be used for clustering, and you would typically try to calibrate them in such a way that a cluster has an average of not more than 200–500 records in them. To deal with fuzziness in your data, you might set up a cluster like using the phonetic last name representation and the same zip code in which case the phonetic variations of *Müller* such as *Mueller* would be in the same cluster, too.

When you implement MDM, usually 4–6 weeks in the project plan is spent where a technical MDM expert configures a fuzzy probabilistic matching engine deciding on which attributes to use, which weights are supposedly applied to an attribute or group of attributes, what the values for the lower and upper thresholds should be, and so on. With an initial configuration, all data is matched and the results are reviewed with business users to make sure the match results satisfy the business requirements, and then adjustments are done where the results were not good enough yet. Each iteration takes usually 2 weeks, and after two to three iterations, the configuration usually meets business user expectations. Incorrect match results are problematic. For example, if there are too many false negatives, you may miss money laundering occurrences. If you need to be in compliance with AML regulations, this could create noncompliance issues. If you have false positives, it could mean you are merging bank accounts of two people which are in reality different. Hence, it is critical to calibrate the matching engine in an MDM system as best as possible to your data to minimize false negatives and false positives.

The last challenge with matching is the number of tasks created for your data stewardship team. Let us assume you load 20 million records to your MDM system. Let us further assume you would have only 1% duplicate tasks as outcome of the initial bulk match. This would yield 200.000 tasks for your data stewardship team. If you assume a data steward processes 50 to 200 tasks a day, it is obvious that it will take either a very long time until all these tasks are resolved or you need to have a lot of data stewards on your payroll which is costly.

With all these challenges, the obvious question is how can AI help to make matching and data stewardship better? We will show the use of AI for three use cases:

1.  Using AI to simplify MDM configuration

2.  Using AI capabilities within the matching engine

3.  Using AI capabilities to significantly reduce the work for data stewards

Let us start with the first use case as shown in Figure 9-5. As mentioned, the initial data load of several sources or in subsequent deployment phases adding more data sources to MDM is a painful and cumbersome process. This task is in fact the long pole in an MDM project, which cannot be substantially shortened neither via additional resources nor via traditional ETL-based approaches. As a result, the time to value is not ideal. Using AI-infused capabilities, this becomes substantially easier. A data engineer using the configuration UX of the MDM system can easily point to one or multiple sources. Also note that the MDM system has been registered in the data governance catalog, and for all attributes, data classes and business terms are assigned.

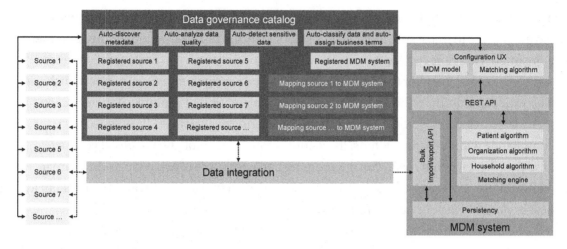

**Figure 9-5.** *Using AI to Simplify MDM Configuration*

Through REST APIs from the data governance catalog, the configuration UX uses all the AI-infused capabilities we previously introduced in Chapter 8, *"AI and Governance,"* which includes

- Auto-discovery of metadata

- Auto-analyze data quality

- Auto-detect sensitive data such as PII

- Auto-classify data and auto-assign business terms

As a result, in the configuration UX, the discovered source is automatically mapped to the MDM data model of the MDM system. To achieve this, the data classification results of the source are compared to the data classes assigned to the MDM data model. Where matches are found, the mapping is proposed as perfect match, otherwise depending on similarity a suggestion is made. The data engineer needs to review the proposed mapping to the MDM data model and either approve it right away or adjust it where needed. Each source becomes a registered source in the data governance catalog as shown in Figure 9-5, and the mappings from each source to the MDM data model are pushed into the data governance catalog, too.

This level of automation removes a huge amount of manual steps and the many UX switches between disjoint data profiling tools, data governance catalog tools, Microsoft Excel – which was often used for the source to MDM mappings – and the MDM tool. Finally, based on AI, one or multiple matching algorithms are proposed depending on the data from the data sources. Again, the data engineer primarily needs to review the proposed algorithms and, if required, might make some minor adjustments. Once that step is done, the one or multiple proposed matching algorithms can be deployed into the MDM system and the initial bulk match can be triggered.

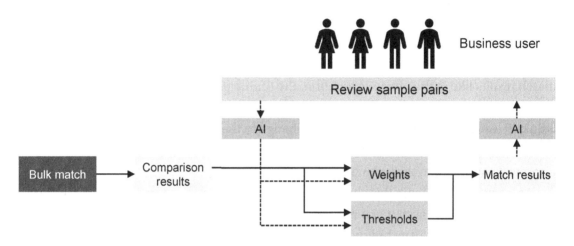

***Figure 9-6.*** *AI-Infused Tuning of the Matching Engine*

With that step completed, we get into our next use case. As previously mentioned, tuning a best in breed probabilistic matching engine was a cumbersome task in the past. With the help of AI, this is significantly changing, bringing us to our second use case shown in Figure 9-6. When the matching engine executes a matching algorithm, for each attribute based on fuzzy matching operators, comparison results are produced. Now, the use of AI allows us to automatically propose an initial set of weights to score the comparison results as well as an initial set of thresholds for the lower and upper thresholds, setting the boundaries for non-match, clerical, and auto-match results. Based on the weights and the thresholds for the initial bulk match, the initial match results are produced.

For example, if you loaded 400 million master data records into the MDM system, a match result is produced for each record in the bulk match process. To verify that the matching configuration meets the business needs, business users can see, for instance, 500–1500 relevant examples – identified with the help of AI – as part of the sample pair review process. Relevant means some of them will be close to each side of the two thresholds to show the business users representative examples for non-match vs. clerical and clerical vs. auto-match. Other examples might represent large clusters of similar results. The business users can then decide for each sample pair if they agree or disagree with the match results, basically creating a truth set through this labeling process. Once all sample pairs are labeled, AI is used again to recalibrate the weights and the thresholds.

You do not need to rerun the matching process, which is time consuming for bulk match on large data sets. With the adjusted weights and thresholds, you can immediately recompute the match results using the stored comparison results of the similarity assessments of the attribute values. If the adjusted weights and thresholds yield identical match results like in the previous iteration, the tuning process ends. Otherwise, the business users need to review and validate the subset of the sample pairs where the adjusted weights and thresholds caused a change in the match results. With this process, the tuning of the matching engine is completed much faster, and the match accuracy is improved compared to the manual tuning without AI.

As introduced earlier in this chapter in the context of Figure 9-1, many MDM solutions use graph capabilities as part of the solution stack. While fuzzy probabilistic matching is certainly a very powerful technique to determine if records are duplicates of each other, it is blind to insights based on the graph representations of master data. A graph is basically a set of vertices with edges connecting vertices. Master data can be represented by modeling records as vertices and relationships between records as edges in the graph. Now take a look at Figure 9-7. The fuzzy probabilistic matching would be able to determine that for the vertices of Robert Smith and Bob Smith, there seems to be some similarity (e.g., first name looks similar based on nickname resolution where "Bob" is a known nickname for "Robert" and on the date of birth there is an edit distance of 1 due to a typo).

However, what is not seen by the fuzzy probabilistic matching is the relationship network around both vertices. That relationship network as shown in Figure 9-8 which is compared for similarity with a reasonable localized scope such as only looking at neighboring vertices which are a limited number of edges[2] away is pretty much the same with the exception of the relationship to the organization vertex "MyComp." In the example, to decide whether or not the two vertices representing the records of Robert and Bob Smith should be merged, the local subgraph around them holds critical

---

[2]First-degree neighbors are all the vertices which are connected with an edge to the records which are compared for similarity. Second-degree neighbors are all nodes which can be reached by crossing at most two edges from the vertices which are compared, and so on. So the local subgraph of degree n is basically the subgraph around each vertex which is compared where we only consider the vertices and edges which are reached in n or less steps.

information for that decision. As you can see, both vertices have relationships to the same wife and the same kids. This relationship information is a very strong indicator that these two records represented by these two vertices should be merged.

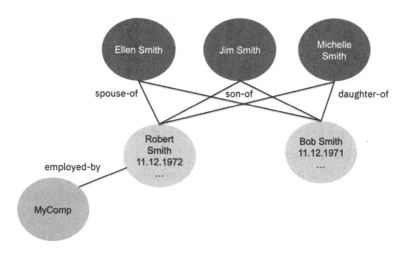

***Figure 9-7.*** *A Graph Example*

With that insight, there are two questions:

1. How can graph representations be used to improve matching?

2. Which AI techniques on graphs improve master data management or drive other analytical useful insights where master data is a key source?

Answering the first question, a number of approaches have been studied[3] on local subgraph similarity measures. Using one of these techniques as an additional feature vector with appropriate weight could be added to combine fuzzy probabilistic matching with a local subgraph similarity measure to improve match results by leveraging the graph structure as well.

For the second question, one example is the graph neural networks (GNN)[4] which have been applied for use cases such as link prediction which basically means to find relationships which are not explicitly declared in the graph data. Using GNN for link prediction within an MDM system allows, for example, to find hidden relationships

---

[3]See, for example, [5], [6], [7], and [8] if you want to learn more about this topic.

[4]The boundaries of what can be achieved with graph neural networks are discussed in this research paper [9].

which might be critical insights in detecting fraud or other malicious activities. GNN for link prediction can be implemented, for example, using open source python libraries such as PyTorch[5]. Customers are currently looking for enhancements by the MDM vendors bringing such graph-based AI techniques into their MDM software solutions. However, if link prediction based on GNN is added to MDM, the explainability of such a prediction needs to be there at the same time. While the explainability of an AI-based prediction is easier for some ML techniques, it tends to be harder for predictions using neural network approaches. At the time of this writing, research is actively looking into the explainability of GNN-based link predictions. We expect in the next few years more GNN-based capabilities to appear within commercial MDM software solutions once the explainability angle is solved as well.

The next use case is the AI-infused data stewardship use case.[6] MDM software like IBM MDM[7] provides AI capabilities to significantly reduce the amount of clerical tasks the data stewardship team needs to process.

Figure 9-8 shows a conceptual overview of the MDM system with key components involved. The key idea is that an ML algorithm is used to learn from the data steward and their decisions while processing the duplicate suspect tasks known as clericals. Then a trained AI model is deployed resolving most of the future tasks automatically. The MDM system is used and the data stewardship team needs to process duplicate tasks for a while. This creates a resolution history of duplicate suspect tasks for the clericals processed. Depending on the ML algorithm used and whether or not optimized training approaches are applied, a different amount of clericals need to be processed before the training can start.

For example, using the ML algorithm random forest, around 5000 resolved clerical tasks were necessary before the trained model produced reliable results. With iterative training where in each training cycle clustering techniques were used to select the next clerical tasks for the data steward to resolve before the random forest model was retrained, the number of resolved clerical tasks could be reduced from 5000 to 250 to 300 tasks.

---

[5]It's the torch_geometric package in PyTorch python package. For more details, see [10].

[6]A scientific research paper on this subject with details of various ML techniques applied including their prediction quality comparison and performance considerations can be found here [11]. It shows that random forest and extreme gradient boosting are comparable relative to prediction quality, but the random forest is much faster to train and requires less hardware resources for training.

[7]See [12] for more information.

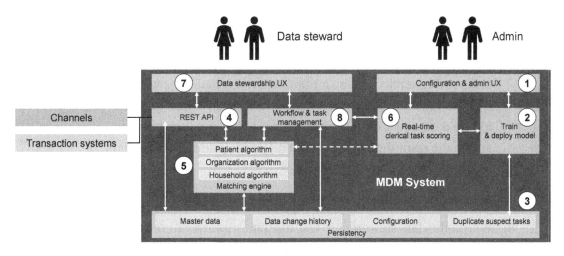

***Figure 9-8.*** *AI-Based Data Stewardship*

This was due to an accelerated beneficial training effect for the random forest algorithm in this approach. In general, the approach works as follows assuming your MDM solution has this capability built in out of the box:

1. Once enough clerical tasks are processed, an admin user can log in to the configuration and admin UX.

2. With the configuration and admin UX, the admin user can train and deploy an ML model. With this fully integrated capability, the UX shows the accuracy of the trained model. Once satisfied, the ML model gets deployed as a real-time scoring service.

3. The data needed for the training of the ML model is available from the resolution history of the processed clericals which are stored in the persistency in the duplicate suspect task area.

4. Channel or transactional applications invoke the REST API from the MDM system to create or update master data records.

5. For create or update operations, the REST API invokes the MDM matching engine with one or multiple algorithms deployed and at least one invoked to perform a matching operation.

6.   In case a clerical task is found, the matching engine invokes the
     real-time clerical task scoring service. Based on a configurable
     prediction quality threshold, the recommendation of the clerical
     task scoring service can be automatically applied if the accuracy
     is above the configured threshold. In this case, no task is added
     for the data stewards. If the accuracy is below the threshold, no
     automatic resolution is executed and a task is added for the data
     stewards to be processed.

7.   A data steward can log in to the data stewardship UX.

8.   Within the data stewardship UX, the data steward can see the
     clerical tasks in the inbox and resolve them using the duplicate
     suspect process workflows. Once a task is picked from the task list,
     the task is scored again since the underlying records might have
     changed since the task was originally created and the data steward
     can see the recommendation of the clerical task scoring service.
     Assuming the data steward agrees, the proposed recommendation
     can be applied with one click. Otherwise, the data steward can
     resolve the clerical task by following the duplicate suspect process
     step by step.

The value proposition is huge. First of all, a significantly smaller amount of clerical
tasks is created for the data stewards and this could be more than a 50% reduction. In
addition, in case the ML prediction is not good enough for auto-resolution, the data
steward can see the proposed recommendation and follow the resolution in parts or
as a whole, depending on his judgment. The prediction quality threshold is initially
configured very high until the data stewards get comfortable with the AI within the MDM
system. Reports of the AI resolved cases as well as reports showing how many of the in
doubt recommendations the data stewards agreed to over time help to more accurately
set the prediction quality threshold. Lastly, features to detect model thrift alerting the
admin that a potential retraining is required complete the AI feature set for the AI-
infused data stewardship capability.

In case your MDM software does not have such a capability built in out of the box,
you can still achieve similar benefits by running a DS project to figure out which ML
algorithm works best in the context of your MDM software and custom integrate the
real-time scoring service into the API layer and the data stewardship UX.

# Operationalizing Customer Insight via MDM

MDM systems as shown in Figure 9-9 have been deeply integrated into the operational IT fabric. Depending on the industry, MDM is serving customer data to the core transactional systems, for example, in banking to systems running credit card, checking and saving accounts, mortgage, and wealth management.

In addition, MDM is providing the customer data to all systems of engagements comprised of the direct and indirect channels. For example, direct channels are mobile or online channels. MDM also serves the indirect channels such as customer care platforms, for example, call centers, CDPs, or marketing automation tools.

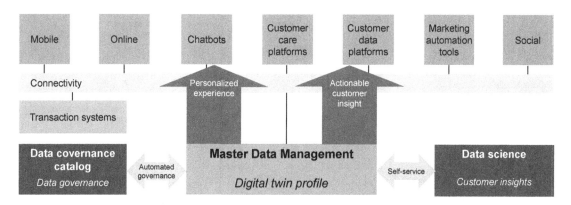

***Figure 9-9.*** *Personalized Customer Experience and Actionable Customer Insight*

Many data science projects are focused on driving deeper customer insights which can be derived on data science platforms. Here are some examples of customer insights:

- Customer segmentation

- Next best action and next best offer

- Product mentions on social media channels

- Topic and sentiment detection

- Churn indicators

- Fraud risk indicators

Data science projects for customer insights suffer from two major shortcomings. First, data scientists oftentimes do not find master data through the shop for data experience in the data governance catalog because the MDM system and the data governance catalog have not been integrated and the MDM system is thus an unregistered data source. As a result, data scientists often spend a lot of time to curate the master data sets they need from the sources with less than optimal data quality even though it might sit in a well-curated but not discoverable MDM system.

Even if the MDM system is a registered information asset in the data governance catalog, it might lack the self-service access for the data science community to search, select, and export relevant subsets of master data into the data science environment.

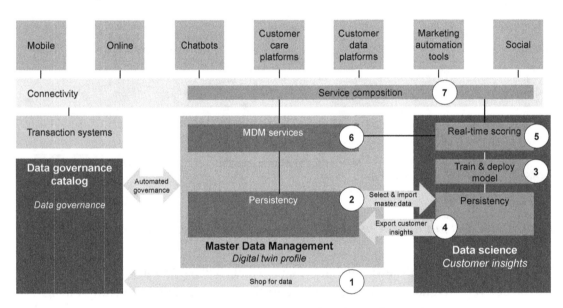

***Figure 9-10.*** *Operationalizing Customer Insight Through MDM*

The second question is how customer insight can be operationalized. Figure 9-10 shows how this can be achieved by integrating the MDM system into the data governance catalog and registering it as a governed data asset.

Given that MDM is already connected to many of the critical places where the customer data is required, expanding the 360° view of the customer in MDM to the digital twin profile within MDM makes sense because of two key business benefits. It allows to correctly engage with a prospect or customer because in MDM you have all the engagement preferences and summary interactions available. In addition, the customer insights become actionable[8]. For example, with the next best action or the next best offer showing up on the screen, your call center agent, your digital sales representative, or your insurance agent can interact with the customer in the most optimal way. All these aspects together deliver a truly personalized customer experience.

Now the following is possible:

1. **Step 1**: A data scientist searches for master data using the shop for data experience and finds the master data in the MDM system.

2. **Step 2**: Assuming the data scientist has the access privileges for master data, the data scientist can now use the MDM UX to search and explore which master data set ideally supports the data science project. For example, maybe the data scientist needs a particular customer data set where all customers are in a certain age range, in a certain region, and all belong to the silver customer segment. Once the data set required is identified, the data scientist uses the MDM bulk export function to export it into the persistency where the data scientist executes the data science analysis.

3. **Step 3**: Using the data science tools, the data scientist trains a new AI model.

4. **Step 4**: If required, the data scientist also performs a bulk scoring of the customer data which can then be imported back to MDM using the bulk import capabilities.

5. **Step 5**: The trained model is deployed as a real-time scoring service on an appropriate AI runtime. The real-time scoring service could be made accessible through an easy-to-use REST API endpoint.

6. **Step 6**: The real-time scoring service could then be integrated into the MDM services layer so that if new customer records are added or existing records get updated, the ML model in the scoring service can derive the required insights in real time which get persisted alongside the other attributes in the persistency of the MDM system.

---

[8]Please review Chapter 5, "*From Data to Predictions to Optimal Actions.*"

7. **Step 7**: Alternatively, using service composition in the connectivity layer, the MDM services and the real-time scoring service can be composed into a single service with the same result as outlined in step 6.

More efficient customer insights projects as well as actionable customer insights become a reality with this deep integration between the data governance catalog, the MDM system, and the data science platform.

# Key Takeaways

As usual, we conclude this chapter with several key takeaways. MDM is around since two decades with broad market adoption since many years. With a fast-growing number of regulations as well as an explosion of data in terms of volume and variety, data governance processes and tools had to be infused with AI to still remain practical, and we showed in this chapter how this is beneficial to MDM as well. Similarly, MDM has to evolve by broadly adopting AI to allow much faster integration of new sources or to keep the workload for data stewards manageable.

Table 9-1 summarizes some of the key insights of this chapter.

***Table 9-1.*** *Key Takeaways*

| # | Key Takeaway | High-Level Description |
|---|---|---|
| 1 | AI radically simplifies the configuration of best in breed probabilistic, fuzzy matching engines | Time to value is much faster if AI-based techniques on auto-mapping a new source to MDM are applied and a configuration of the matching engine is proposed based on AI. Accuracy of the match results is also improved with using AI for weight and threshold assignments |
| 2 | AI reduces the labor of data stewards processing duplicate clerical tasks in MDM systems by 50% or more | Learning from the resolution history of clerical tasks, real-time scoring of new clerical tasks based on trained ML models radically reduces the effort for data stewards. This is a major step forward to improve data quality, business outcomes, and cost reduction. Unresolved clerical tasks could represent a missed AML case causing noncompliance issues and customer churn since the same person still receives multiple marketing materials |

*(continued)*

***Table 9-1.*** (*continued*)

| # | Key Takeaway | High-Level Description |
|---|---|---|
| 3 | Operationalize customer insights through MDM to make them actionable | MDM systems are connected to many systems of engagement and transactional systems. By integrating your DS platform with MDM, you can operationalize your AI-derived customer insights through MDM, delivering a true personalized customer experience as well as making customer insights actionable when customers interact with your enterprise |

# References

[1]   Allen Dreibelbis, Eberhard Hechler, Ivan Milman, Martin Oberhofer, Paul van Run, Dan Wolfson: *Enterprise Master Data Management. An SOA Approach to Managing Core Information.* ISBN-13: 978-0134857503, Pearson Education, 2018.

[2]   Mark Allen, Dalton Cervo: *Multi-Domain Master Data Management: Advanced MDM and Data Governance in Practice.* ISBN-13: 978-0128008355, Morgen Kaufmann, 2015.

[3]   Eberhard Hechler, Ivan Milman, Martin Oberhofer, Scott Schumacher, Dan Wolfson: *Beyond Big Data. Using Social MDM to Drive Deep Customer Insight.* ASIN: B00OM1MBKA, IBM Press, 2014.

[4]   Alex Berson, Larry Dubov: *Master Data Management and Data Governance.* ISBN-13: 978-0071744584, McGraw-Hill Education, 2000.

[5]   Emilio Ferrara, Palash Goyal: *Graph Embedding Techniques, Applications and Performance: A Survey.* 2017 `https://arxiv.org/pdf/1705.02801.pdf` (accessed May 2020).

[6]   Vincenzo Carletti: *Exact and Inexact Methods for Graph Similarity in Structural Pattern Recognition.* 2016. `https://hal.archives-ouvertes.fr/tel-01315389` (accessed May 2020).

[7]   Cheng Deng, Weiyao Lin, Junchi Yan, Xu-Cheng Yin, Hongyuan Zha, Xiaokang Yang: *A Short Survey of Recent Advances in Graph Matching.* 2016. `www.researchgate.net/publication/303901965_A_Short_Survey_of_Recent_Advances_in_Graph_Matching` (accessed May 2020).

[8]   Sergey Melnik, Hector Garcia-Molina, Erhard Rahm: *Similarity Flooding: A Versatile Graph Matching Algorithm and its Application to Schema Matching.* In: IEEL. 18th International Conference on Data Engineering. [S.l.], 2002. `http://ilpubs.stanford.edu:8090/730/1/2002-1.pdf` *(accessed May 2020).*

[9]   Stefanie Jegelka, Jure Leskovec, Weihua Hu, Keyulu Xu: *How Powerful are Graph Neural Networks?* In: Proceedings of the ICLR 2019, May 6-9, 2019. `https://arxiv.org/abs/1810.00826` (accessed May 2020).

[10]   PyTorch: `https://pytorch.org/` (accessed May 2020).

[11]   Lars Bremer, Mariya Chkalova, Martin Oberhofer: *Machine Learning Applied to the Clerical Task Management Problem in Master Data Management Systems.* In: BTW 2019, p, 419–431.

[12]   IBM Master Data Management: `www.ibm.com/products/ibm-infosphere-master-data-management` (accessed May 2020).

# CHAPTER 10

# AI and Change Management

As AI is increasingly adopted by businesses and society as a whole, there is an emerging need for existing change management practices to be adapted. Change is usually perceived as a threat, causing uncertainty, sentiments, and risks to organizations and individuals alike. However, change comes along with new business and personal opportunities. AI has the potential to accelerate and improve change management and make it more unerringly and human-centric.

This chapter sheds some light on change management in the context of AI and introduces key aspects of AI change management, such as identifying and analyzing sentiments for a more targeted change management with optimized outcome. We also elaborate on challenges to incorporate AI into change management and highlight the importance of data- and information-driven aspects to embrace AI for an increased insight-driven and pervasive change management approach.

## Introduction

AI will have a profound impact on how businesses operate, how projects are executed, and how people work together, for instance, using chatbots and robot devices. On one side, AI techniques represent a great opportunity for change management in that AI can be leveraged to improve and augment traditional change management methods and tools, for instance, by applying ML algorithms to predict project outcomes with higher precision, or to discover unknown risks that may cause project changes.

© Eberhard Hechler, Martin Oberhofer, Thomas Schaeck 2020
E. Hechler et al., *Deploying AI in the Enterprise*, https://doi.org/10.1007/978-1-4842-6206-1_10

On the other side, change management is facing challenges to accommodate AI, meaning to bear in mind the impact that AI has on organizations, projects, and individuals, for instance, in the way AI and ML will accelerate change of the organizational structure, business models and processes, and the way people work and learn. This has to be anticipated and taken into consideration in terms of adjusting relevant change management practices in your organization. This impact of AI on organizations, projects, and individuals requires a degree of adjustment by the existing change management discipline that remains unsurpassed by any prior technology impact.

In this chapter, we focus on AI-driven improvements for change management, for instance, to leverage AI and ML to optimize IT change management or to advance architecture change management. These are more technical-oriented topics. In addition, we also elaborate on project and human resource (HR)-related aspects, such as applying ML and DL methods to infuse social media-based sentiment analytical insight into change management, and the influence of AI and ML on the well-known and essential project management discipline, including the usage of AI and ML in HR management to streamline the change management process regarding resource allocations.

# Scope of Change Management

This section revisits traditional change management and provides a definition as well as a high-level overview of change management models, processes, and approaches.

## Change Management – Scope and Definition

Change management is usually seen as a discipline of project management. However, limiting change management in such a way may be too narrow-minded, since change isn't only occurring while executing projects. Especially approaching change management in the context of AI doesn't live up to expectations if it is strictly limited to the project management domain.

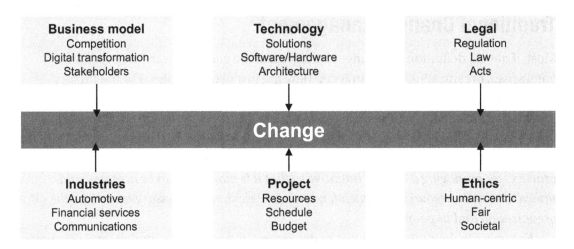

**Figure 10-1.** *Reasons Causing Change*

Change management needs to be driven by clearly stating business objectives. More often than not, change is conducted just for the sake of it. The availability of change management tools amplifies this trend especially in customer data-related projects, where IT vendors provide tools that often mislead organizations to approach changes with missing or unclear business objectives.

Reasons for change are more pervasive; change itself is accelerating and is more profound, particularly if driven by AI and ML and new business scenarios. Complementing project-related changes, we need to factor in new legal and ethical aspects, change driven by industry-specific scenarios (e.g., self-driving cars) and new technologies (e.g., AI and ML), and emerging business models. Figure 10-1 illustrates some key reasons for change and provides the foundation for a more adequate scope and definition of change management to be realized by organizations, as well as individuals and the society as a whole. The referenced industries in Figure 10-1 are just some examples.

With simultaneous consideration of these drivers of change, we suggest the definition of change management as an organized, systematic approach to apply knowledge, tools, processes, models, techniques, and resources resulting in the adoption and realization of change. For organizations, this means to achieve their business goals and strategy in the light of change. For individuals, this means to simply accept and adopt new ways of communicating (e.g., via chatbots), commuting (e.g., via self-driving cars), or learning (e.g., via personalized recommended education). For the society, this means to adopt change by developing, for instance, new ethical, human-centric guidelines. As you can certainly see, AI is setting the pace for the way we need to look at change management.

# Traditional Change Management

Most, if not all, definitions of change management are geared toward corporations, enterprises, organizations, and projects. Here are just two examples: For the Project Management Institute (PMI),[1] for instance, *change management is an organized, systematic application of the knowledge, tools, and resources of change that provides organizations with a key process to achieve their business strategy.* The Society for Human Resource Management (SHRM)[2] defines change management as the *principles and practices for managing a change initiative so that it is more likely to be accepted and provided with the resources (financial, human, physical, etc.) necessary to reshape the organization and its people.*

Not only the definition of change management, the approach, models, and processes of change management are centered around organizational and project managerial aspects.[3] The traditional change management concept won't lose its validity; however, we need to develop novel ideas for an approach to change management that takes into consideration the profound impact driven by AI. In the past, change management always faced challenges that were related to resistance and hesitance for change and the tendency to drop back into well-known patterns and ways of conducting business. With AI penetrating the society as a whole and its impact on businesses and individuals alike, hesitation and skepticism, rejection and fear, and even gaps in knowledge and skills impose the creation of innovative ideas and a paradigm shift toward traditional change management.

Because of space constraints regarding this chapter, and since change management is just one particular topic to be discussed in this book, we won't be able to exhaustively outline an AI change management framework. However, this chapter pursues the goal to at least introduce the key ideas and cornerstones to develop such a framework.

# Change Management in the Context of AI

In this section, we continue to discuss the influence of AI on change management and also list some of the important and pivotal challenges that change management is going to be faced with. Change management has always been and will continue to be information driven (including data). Because of AI, you can derive new insights, which is of relevance to make change management more precise and targeted.

---

[1]See [1] for more information on the PMI definition of change management.
[2]See [2] for more information on the SHRM definition of change management.
[3]See [3] for a comprehensive treatment on leading change.

In this section, we therefore discuss the extension that AI facilitates by accelerated inference of new insights, which can then be leveraged by change management. We especially highlight the impact that AI has on change management and provide an example on how AI is driving change on organizational structures regarding HR, skills and knowledge, and education with new ways of learning.

## AI Influence on Change Management

In order to better understand the influence of AI on change management, we identify the key elements or components of traditional change management. These elements or components will be examined regarding their required adjustment and expansion initiated by AI and its siblings. In pursuing this approach, we should be cognizant in regard to the possibility or even necessity to add new components.

There are literally hundreds of books[4] available on change management models and methodologies, frameworks and processes, tools, and other aspects. In this chapter, we need to be selective and focused. We therefore limit our approach to the following well-established elements and components of traditional change management:

1. **Change management models and methodologies**: There are numerous change management models and methodologies described. Most of them are geared toward organizational and project interests[5]; some of them are complex and cumbersome and lack organizational change aspects. AI calls for models and methodologies to be augmented to collectively embrace individuals and groups, consumers and users, and even society as a whole. Enterprises and organizations need to broaden the models and methodologies beyond their immediate boundaries of projects, organizations, teams, and employees and adapt a leadership role far beyond their comfort zone.

2. **Change management processes**: Change management processes include a well-defined sequence of activities or tasks that transform a change from its occurrence to successful consideration in regard to defined goals. If you recall the reasons

---

[4]See [4] for more information on the theory and practice of change management.
[5]See [5] for more information on change management models and methodologies.

for change as described in Figure 10-1, it becomes obvious that AI-driven change management processes have to include the additional steps that take into consideration particularly legal aspects, ethics, and AI technology itself.

3.  **Change management KPIs and metrics**: Taking AI into consideration, relevant change management KPIs and metrics may have to be adjusted; additional ones can and should be added. For instance, AI and ML may be used to automatically discover unfavorable deviations from your project baseline or to detect critical sentiments from outside your organizational boundaries derived from social media analytics[6].

4.  **Change management tools**: There are quite a number of change management relevant tools, including governance tools, document management and communication tools, issues and problem management tools, and project management and scheduling tools – to just mention a few. AI with ML capabilities and data science methods can and should enrich these tools to increase precision and accurateness, to generate more relevant and targeted insight, and to predict and automatically discover and even suggest preventive actions.

5.  **Change management culture**: Adapting the change management culture by truly embracing AI and leading change requires an open, agile, and proactive attitude. This affects the corporate and organizational culture, as well as the cultural setting of individuals, communities, and the society as a whole. AI will influence the change management culture. For instance, ML and data science methods can be used to understand sentiments and propose targeted and customized actions.

6.  **People and human aspects**: AI already has a strong impact on stakeholders, employees, individuals and citizens, communities, and the society. Change management was led and driven by people. AI may change this paradigm: people may get assistance

---

[6]See [6] and [7] for more information on social media analytics.

through AI-based advisory functionality; in some way, they may even get replaced with AI-powered change management tools sitting in the driver seat.

7. **Information** (including data): Change management is strongly information driven. The insights derived from data and information fuel the change management processes, including recommended actions. AI intensifies this and makes insight more predictive and relevant and derived actions for adjustments more targeted and customized.

In addition to the preceding examples, additional AI-related aspects that are impacting change management can be derived from the challenges, which we describe in the following section.

# Challenges for Change Management

There are numerous existing challenges for change management that become more intense and prevalent through AI; new ones arise as well. Table 10-1 lists the most important challenges. In a way, most of these are derived from our understanding of change as outlined in Figure 10-1.

*Table 10-1.* *Challenges for Change Management*

| # | Challenge | Description |
|---|-----------|-------------|
| 1 | Challenging legal obligations | Regulations will only increase and create further changes and impact on change management to become more agile and responsive |
| 2 | Knowledge and skill gaps | Insufficient knowledge, skills, and understanding about AI technology, scenarios, devices, and tools will make change management more difficult |
| 3 | High complexity | For instance, using AI to improve change management tools requires a level of ease of use and consumability of insight, which needs to hide the AI complexity |

(*continued*)

***Table 10-1.*** (*continued*)

| # | Challenge | Description |
|---|-----------|-------------|
| 4 | Insufficient acceptance | Requires novel ideas regarding communication and information campaigns within change management processes to address concerns by individuals, organizations, and the society |
| 5 | Societal refusal | Refusal by society may cause change and adjustments on goal settings and also impacts change management to take refusal into consideration upfront |
| 6 | Ethical concerns and uncertainty | Change management needs to validate and check ethical concerns, human-centricity, fairness, and benefits for society |

With regard to ethical concerns, there are actually two related aspects: on one side, change management needs to perform the validation and check as described previously; on the other side, AI- and ML-infused change management tools need to adhere to defined ethical guidelines, for example, they need to be fair and human acceptable as well.

These challenges should be taken as a base to derive to a suitable set of success criteria. For instance, insufficient acceptance or societal refusal of the new AI technology needs to be addressed with adequate information campaigns and education programs, leading conversations with worriers, and carefully scaled acceptance trials.

# Driving Change on Organizational Structures

Organizational structures are already significantly influenced by AI; this influence only becomes stronger and more sustainable as organizations are moving up their ladder to AI. Organizations need to become more agile[7] and flexible to adopt and facilitate change more easily. This flexibility and agility relates to the adoption of AI in terms of changing attitudes, user behavior, and often occurring resentments. Dealing with new AI technology, tools, devices, robots, and so on within the organization will lead to organizational adjustments.

---

[7]See [8] for more information on agile organizations.

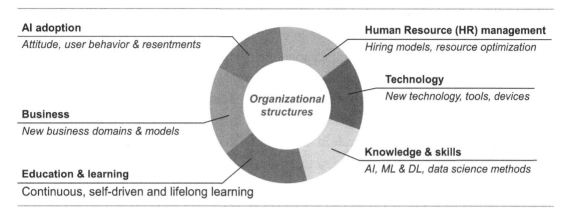

**AI adoption**
*Attitude, user behavior & resentments*

**Human Resource (HR) management**
*Hiring models, resource optimization*

**Technology**
*New technology, tools, devices*

*Organizational structures*

**Business**
New business domains & models

**Knowledge & skills**
*AI, ML & DL, data science methods*

**Education & learning**
Continuous, self-driven and lifelong learning

***Figure 10-2.*** *Impact on Organizational Structures*

As the business changes under the influence of AI, this means for organizations to develop into new business domains and models. Human resource management will continue to change regarding tools and hiring models and resource reallocation and optimization processes. For instance, ML and data science methods will be increasingly used to improve candidate evaluation and mapping to available job roles and responsibilities. This requires cognizant attitudes and careful consideration by HR professionals to guarantee fairness and respectful and transparent decisions.

In the past, educational programs within the organizations have already changed, mainly through the Internet and online learning methods. However, AI will make possible yet again new ways of learning, for example, customized and recommended learning, automated discovery of skill gaps, or strengths in certain areas. Another dimension of organizational changes is imposed by the required knowledge and skills of AI. Although there is no requirement for everyone to become a data scientist and AI subject matter expert, most members of an organization need to become familiar with AI and key data scientist methods.

Figure 10-2 is an illustration of the various driving changes on organizational structures.

# Key Aspects of AI Change Management

Change management should be viewed as *anticipating* and *leading* change, which means to proactively identify opportunities for change that can lead to improved operational efficiency of organizations or communities, lowering risks of projects or failure of technology adoption, capturing and realizing new business opportunities, and optimizing societal and public togetherness (e.g., in commuting, communication, lifestyle).

Adopting AI and ML to improve the efficiency of the change management discipline has many different facets. As we have seen in the previous section, existing tools and methods that are used to facilitate change management can be significantly improved by, for instance, applying predictive analytics, correlation algorithms, and anomaly detection mechanisms to become more precise in predicting a possible budget overrun, to discover correlations between incidents and risks, and to detect project anomalies with higher accuracy and early in the project cycle in order to reduce the degree of impact.

In this section, we provide just a few examples that illustrate the possible enrichment and opportunity that AI represents for change management. For instance, AI and ML can be leveraged to improve the existing IT and architecture change management processes; AI methods can be utilized to improve HR management processes (as we have briefly touched in the previous section as well).

Last but not least, we point out the importance of identifying sentiments portrayed via the numerous social media channels and its relevance on change management.

# AI Change Management Framework

As you can certainly imagine, developing a comprehensive AI change management framework[8] can be a challenging and elaborating task, which we cannot achieve even rudimentarily in just a small section of this book. However, we intend to at least provide some key ideas and design points for such a framework in an AI context. In other words, the goal is to describe what needs to change and where does change occur in a change management framework *because* of AI, so we are depicting influencing changes to an existing change management framework that are caused by AI and its siblings: the AI-causing impact on a change management framework.

---

[8]See [9] for more information on change management frameworks.

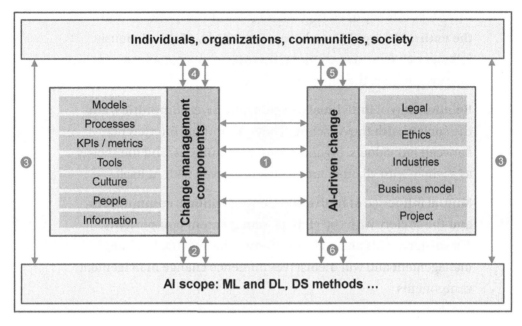

***Figure 10-3.*** *Change Management Framework in the Context of AI*

These ideas can serve as input for the interested reader to progress with a much more refined AI change management framework. We are obviously doing this in the context of what has been described so far in this chapter in terms of AI-driven changes, the key change management components or elements, the impact of AI, and so on.

The correlation and dependencies of change management are multifold. Figure 10-3 is depicting these aspects in their mutual interference. Figure 10-3 depicts four areas or domains: the change management components, the AI-driven change, the AI scope itself, and individuals, organizations, communities, and the society. There exists a mutual interdependency – meaning dependencies in both directions – between the various domains[9].

The following is a short description of these interdependencies:

1. **Mutual interference of change management components and AI-driven change**: Traditional change management components have to be adjusted, for instance, by emerging regulations and ethical concerns or industry-specific needs. Ethics and legal aspects may have a particularly strong impact. The other way around, change management will obviously influence business models and projects.

---

[9]The numbers in the following list correspond to the numbers in Figure 10-3.

2. **Interdependency of change management components and the entire AI scope**: As we have seen already, AI will obviously influence change management; however, change management may even impact the future AI scope itself.

3. **Relationship of individuals, organizations, communities, and the society with the AI scope**: There is a mutual interference between individuals, organizations, communities, and the society as a whole with all other areas, obviously even with AI itself.

4. **Mutual influence of individuals, organizations, communities, and the society with the change management components**: These individuals and communities will be affected by change management and will themselves influence change management components.

5. **Interference of AI-driven change and individuals, organizations, communities, and the society**: AI-driven changes will impact communities, and vice versa, communities will create changes as well.

6. **Relationship of the entire AI scope and the AI-driven change**: Last but not least, AI technology will create the majority of changes; however, the need for business models and projects to be adjusted (because of change requests) may identify AI gaps that will spark new ideas for new, innovative AI capabilities.

This proposed change management framework may be straightforward, even simple; however, it provides a model for further innovative ideas and refinement.

# AI for IT Change Management

As a concrete example regarding the impact of AI on change management, let us have a closer look at IT change management. There are numerous IT change management approaches and tools available. One approach is ITIL[10], which is a set of best practices for managing IT services. IT change management is considered as one of many

---

[10]See [11] for more information on Information Technology Infrastructure Library (ITIL).

processes of the ITIL *service transition* stage. The latest ITIL edition 4[11] has taken new initiatives and themes into consideration, such as digital transformation, lean, agile, and DevOps. There are other frameworks, for example, COBIT[12]; however, ITIL includes a rather detailed process description for IT change management.

We intend to sketch out the impact of AI on IT change management by taking the core steps of a typical IT change management process, outlined in Figure 10-4, as the base. The scope of the changes is leaned toward ITIL best practices; the impact of change can relate to any IT aspect, including SW and HW, applications and solutions, and IT services.

The above outlined IT change management process is a general one, which can be refined and adjusted, given a concrete IT environment. It can also be adjusted to specific organizational needs to implement a more agile process. The impact of AI on the larger IT service management (ITSM) area is obviously much larger than just IT change management. In this section, however, we limit our discussion strictly to the IT change management domain.

The influence of AI on IT change management will be rather impactful; offering new paradigms, which are characterized by data insight-driven decision making and optimization relevant for all process steps. AI will transform IT change management into a subarea of AIOps, for instance, by automating IT change management, recommending adjustments to the IT landscape, service-level improvements, new technical capabilities, architecture building blocks, and products. It enables pattern discovery and matching of a change request (CR) to previous CRs, underpinned by past data exploration, to carbon copy and learn from past experiences and available data. This can be a significant improvement for the evaluation, preparation, and implementation steps of the IT change management process.

---

[11]See [12] for more information on ITIL Foundation edition 4.

[12]See [13] for more information on the Control Objectives for Information and Related Technologies (CODIT).

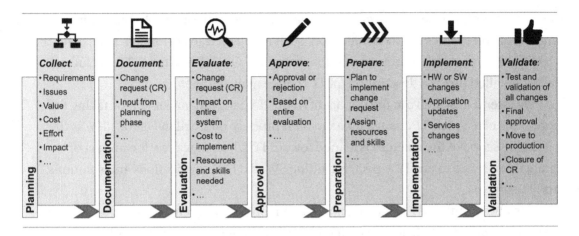

**Figure 10-4.** *IT Change Management Process Steps*

AI enables impact analysis of a CR to discover conflicts or issues caused by a particular request. AI for IT change management may not only discover issues, but has the potential to develop strategies and recommendations on how to efficiently respond to a request. As you have seen in Chapter 8, "*AI and Governance*," AI can be used to identify various risks related to IT change management. It can furthermore autonomously recommend adequate mitigation strategies that can advance the evaluation step of the change management process.

The evaluation and preparation steps can be performed in a more targeted way, for instance, by selecting required products, offerings, tools, and services, including the dependencies and prerequisites among them through the support of AI. During all process steps, issues and problems can be addressed more professionally based on AI-driven analytics of best practices and knowledge related to the IT environment, guidelines and blueprints, and records of past experience available in your organization.

AI can also be used to optimize the workforce, meaning to identify skills and subject matter expertise and other resources needed to manage a CR. Already during the first steps of the IT change management process (planning, documentation, etc.), AI can be used to intelligently analyze project-relevant collaterals, emails, documents, relevant regulations, best practice papers, and so on. AI-based optimization can also be applied to the optimization of the process or workflow itself, turning IT change management into a set of lean and intelligent processes. Change management in general – applicable to IT as well – can be improved via social media analytics, understanding sentiments from users or the public, and taking this into consideration. We explore this further in the following section.

These are just a few concrete ideas on how AI can impact and optimize the IT change management process. However, this requires adaptation in the corresponding IT change management tools.

# Social Media Analytics to Optimize Changes

Change management will be impacted by public sentiments: individuals and users of AI applications, organizations and communities, and even the public itself will actively share comments via forums and social media. This is most likely related to the whole scope of challenges for change management as listed in Table 10-1. These comments and sentiments, resentments, and concerns – although very likely perceived as negative – should be viewed as desirable input to the change management process.

With the many websites, forums, and channels (e.g., Facebook, Twitter, LinkedIn, YouTube, SlideShare, Instagram, Reddit), social media presents a wealth of data to serve for listening (monitoring and quality enhancement objectives), visualization, and analytics to understand and act upon opinions and sentiments, issues and problems, and acceptance or resistance in a holistic and pervasive way.

Social media analytics[13] – as a subdomain of AI – can support desirable adjustments (changes) for the following business domains:

- **Competitive insight**: Gaining insights and comparing your products or services with competitive offerings can drive for competitive advantages.

- **Product and service optimization**: Understanding deficiencies and gaps will drive changes to optimize products, tools, and services.

- **Acceptance improvements:** Sentiment analytics help to discover key acceptance-related issues, preventing sales or other relevant metrics within or outside the organization, for example, ethical concerns.

- **User experience:** Listening to and analyzing social media provides insight on user experience, for example, in using chatbots, self-driving cars, robotics devices, and so on.

---

[13]See [14] for more information on the capabilities of social media analytics.

- **Improving organizational structures:** Social media analytics helps to discover gaps (e.g., knowledge, skills, etc.), suggesting organizational adaptations.

- **IT service management optimization:** Required changes can be derived from social media analytics to further optimize ITSM, for example, improving SLAs.

As can be seen by the preceding domains, social media analytics represents an opportunity to implement change itself and to even adjust the AI change management process.

# Key Takeaways

We conclude this chapter with a few key takeaways, summarized in Table 10-2.

***Table 10-2.*** *Key Takeaways*

| # | Key Takeaway | High-Level Description |
|---|---|---|
| 1 | Many reasons causing change | Take into account the wide scope of reasons causing change: for example, ethics, legal aspects, new AI technology, new business models, and so on |
| 2 | Understand full scope of challenges | Change management challenges may shift; new ones will emerge, for example, societal acceptance or refusal, high complexity, knowledge and skill gaps, and so on |
| 3 | Significant changes on organizational structures | Organizational structures will experience significant changes related to HR (hiring models, etc.), knowledge and skills, education and learning, and so on |
| 4 | Need for an AI change management framework | The AI change management framework will be much broader and determined by multifold correlations and dependencies |
| 5 | IT change management | The IT change management process steps will experience profound enhancements driven by exploitation of AI |
| 6 | Social media analytics | Sentiments, comments, and resentments, derived from AI-based social media analytics should serve as useful input to change management |

# References

[1]   PMI. *Integrated change management.* `www.pmi.org/learning/`
      `library/integrated-change-management-5954` (accessed
      September 14, 2019).

[2]   SHRM. *The SHRM Body of Competency and Knowledge.* `www.`
      `shrm.org/certification/Documents/SHRM-BoCK-FINAL.pdf`
      (accessed September 14, 2019).

[3]   Kotter, J.P. *Leading Change.* ISBN-13: 978-1422186435, Harvard
      Business Review Press, 2012.

[4]   Hayes, J. *The Theory and Practice of Change Management.* ISBN-
      13: 978-1352001235, Red Globe Press, 2018.

[5]   Smartsheet. Which (of the Numerous) Change Management
      Models and Methodologies is Right for Your Organization. `www.`
      `smartsheet.com/which-numerous-change-management-models-`
      `and-methodologies-right-your-organization` (accessed,
      September 18, 2019).

[6]   PMI. *Integrated change management.* `www.pmi.org/learning/`
      `library/integrated-change-management-5954` (accessed
      September 19, 2019).

[7]   IBM. Social media analytics – Uncover insights in social media
      to help your business. `www.ibm.com/topics/social-media-`
      `analytics` (accessed September 21, 2019).

[8]   IBM. Tone Analyzer – Understand emotions and communication
      style in text. `www.ibm.com/watson/services/tone-analyzer/`
      (accessed September 23, 2019).

[9]   Kreutzer, R.T., Neugebauer, T. *Digital Business Leadership: Digital
      Transformation, Business Model Innovation, Agile Organization,
      Change Management (Management for Professionals).* ISBN-13:
      978-3662565476, Springer, 2018.

[10]    Franklin, M. *Agile Change Management: A Practical Framework for Successful Change Planning and Implementation.* ISBN-13: 978-0749470982, Kogan Page, 2014.

[11]    AXELOS. *What is ITIL Best Practice?* www.axelos.com/best-practice-solutions/itil/what-is-itil (Accessed September 25, 2019).

[12]    ALEXOS. *ITIL Foundation – ITIL 4 Edition.* ISBN-13: 978-0113316076, The Stationary Office Ltd, 2019.

[13]    ISACA. COBIT 4.1: Framework for IT Governance and Control. www.isaca.org/Knowledge-Center/COBIT/Pages/Overview.aspx (accessed September 27, 2019).

[14]    IBM. Business analytics blog - Getting to know Watson Analytics for Social Media capabilities: Conversation clusters and more. www.ibm.com/blogs/business-analytics/watson-analytics-for-social-media-capabilities-conversation-clusters/ (accessed September 27, 2019).

# CHAPTER 11

# AI and Blockchain

Most people believe that the paper from 2008[1] by Satoshi Nakamoto, a pseudonym used by a yet unknown author, introduced the concept of blockchain. However, the key idea is actually 17 years older. The first mentioning of key blockchain concepts goes back to 1991[2] when Stuart Haber and Scott Stornetta described the concept of a cryptographically secured chain of blocks for the first time.

Blockchain is at its core an immutable, shared ledger application which can be used to record transactions of assets. Assets can be tangible assets of the real world like containers, cars, houses, and many more or digital assets like currencies. Any blockchain technology today employs a couple of key concepts. First, there is the smart contract which is a piece of code that encapsulates the business terms executed within transactions which are recorded by the blockchain. The second key concept is the shared ledger which is conceptually a distributed database holding the records of the recorded transactions. The records are created by multiple participants in a peer-to-peer network. The peer-to-peer networks can be public or private. Each record goes into a block. The block is linked to its predecessor and successor block in a linked list structure – hence the name *blockchain* as illustrated in Figure 11-1.

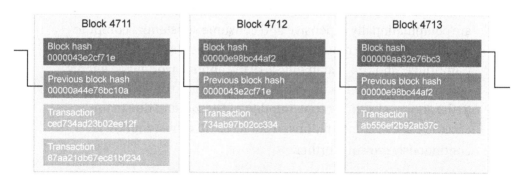

***Figure 11-1.*** *Blockchain*

[1]Here you can find the Bitcoin-related paper by Satoshi Nakamoto [1].
[2]This is the research paper by Stuart Haber and Scott Scornetta [2].

© Eberhard Hechler, Martin Oberhofer, Thomas Schaeck 2020
E. Hechler et al., *Deploying AI in the Enterprise*, https://doi.org/10.1007/978-1-4842-6206-1_11

Each block is timestamped and cryptographically secured. The blocks are immutable and cannot be changed once created by a participant in the blockchain peer network. A key difference between blockchain technologies and the well-known database technologies is the following: A blockchain supports only two operations. Create a new transaction and the read operation of a recorded transaction. As a result, think of blockchain technologies as an append-only persistency. Database technologies have four operations: creation of a record, update of a record, deletion of a record, and a reading of a record (collectively referred to as CRUD operations). The mechanism to prevent that a single participant can change something is done through consensus mechanisms. There are different techniques used to implement consensus such as Practical Byzantine Fault Tolerance (PBFT) or multi-signature. The chain is stored at every participant's node. A real-time synchronization keeps the individual peers synchronized by the blockchain fabric at all times. Strong cryptographic algorithms are used to enforce that all transactions are secure, authenticated, and verifiable. In permissioned blockchains, they are also used to ensure that participants can only access those parts of the ledger they are entitled to see.

Today, the blockchain technology space can be roughly divided into two major categories:

- **Cryptocurrencies**: The first one in this category is Bitcoin; many others like Ethereum and Ripple[3] followed, and there are hundreds of different cryptocurrencies today. Cryptocurrencies are permissionless, public ledgers. The key idea of cryptocurrencies is that they represent a digital currency without bills and coins which you can use to pay for goods when making online purchases. A good introduction to cryptocurrencies can be found here.[4] In public blockchains like Bitcoin where participants are anonymous, consensus is reached through proof-of-work mechanisms, which are computationally expensive and consume substantial compute resources and huge amounts of electricity. The University of Cambridge[5] showed that in July 2019 the annual, worldwide energy consumption of the Bitcoin blockchain was approximately equal to the annual electricity consumption of Switzerland due to the excessive use of computational resources. This consumption is continuously growing further.

---

[3]More details on Bitcoin, Ethereum, and Ripple can be found here [3], [4], and [5].

[4]The key concepts on how cryptocurrencies work can be found here [6].

[5]The University of Cambridge has a tool showing the consumption of electricity of the Bitcoin blockchain located here [7].

- **Permissioned blockchains for shared ledger applications**: This
  type of blockchain technology is used in the enterprise space where
  usually two or more enterprises collaborate. Hyperledger[6] blockchain
  is one of the leading technology examples in this space. It is an open
  source effort hosted by The Linux Foundation and supported by
  a very large number of companies like Accenture, IBM, and many
  others. Hyperledger blockchain is comprised of several projects.
  Permissioned blockchains like Hyperledger where the identities
  of the participants are known can deploy consensus mechanisms
  which are substantially more environmentally friendly, using less
  computational resources and electricity.

Both types of blockchain technologies are recording transactions. However, transactions created through permissioned blockchains which are used more broadly in the enterprise space are typically of more interest to enterprises from an analytical standpoint where AI comes into play. In this chapter, we explore the concept of permissioned blockchains in the enterprise space and look at why companies are implementing blockchains. Afterward, we show how to apply AI techniques to blockchain data. In the scientific research space, there are also many research initiatives at the intersection of blockchain and AI still underway.[7] In addition, we will introduce technologies which started to embed blockchain concepts. The key reason for this is that despite all the hype around blockchain technologies, it is currently hard to predict how much adoption the new blockchain technologies are actually getting in the enterprise space vs. how many use cases will be deployed on existing technologies which got some enhancements based on blockchain concepts. In the last section of this chapter, we discuss how blockchain technology can help with AI governance which we introduced in Chapter 8, *"AI and Governance."*

---

[6]You can find Hyperledger here [8].

[7]You can learn more about that in this research paper [9].

# Blockchain for the Enterprise

To understand the value proposition of a blockchain solution for the enterprise, let us take a look at Figure 11-2. There are business scenarios where certain business processes span across multiple participants. Each of them has its own ledger. Between the ledgers are custom build integration interfaces which can be batch, periodic, near real time, or real time. As a result, the data stored in them is rarely if ever in sync (thus the different gray shades for the ledgers). In addition, these ledgers use traditional database technologies which means their security relies on the administrators operating them. If one of them turns into a disgruntled employee or if a skilled attacker comprises their security credentials, the data in these ledgers is at risk. The data in these ledgers usually comprises business transactions, many of them sensitive such as bank account transactions and others. In a nutshell, this approach is inefficient, vulnerable, and costly.

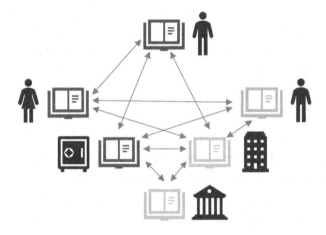

***Figure 11-2.*** *Problems with Traditional Ledgers*

The promise of a blockchain-based solution for enterprises is thus to establish a shared, trusted ledger in a much more effective and efficient way with appropriate security controls in place. In the next section, we introduce the Hyperledger blockchain technology adopted by many enterprises today. Then we show a Hyperledger implementation in the subsequent section.

# Introduction to the Hyperledger Blockchain

Hyperledger is an open source blockchain technology supported by many companies. Enterprises adopt Hyperledger because it provides a shared, replicated, and permissioned ledger with consensus, provenance, immutability, and finality features as shown in Figure 11-3. Each participant is running a peer node of the Hyperledger peer network. The peers are kept in sync through a replication mechanism so that all ledgers of all six participants have the same data at all times. Through the Hyperledger channel and private data collection features, you can control access to the transactions. You can use these features so that, for example, three participants collaborate on one channel, whereas a different group participates on another channel and only see subsets of data they have permissions for. Whether or not that is needed depends on the business processes and participants using the Hyperledger blockchain.

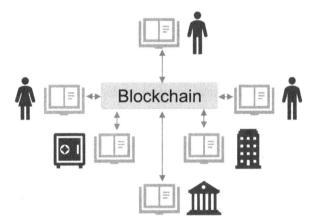

***Figure 11-3.*** *Hyperledger Blockchain*

Hyperledger is comprised of several projects – the following list is only a small subset:

- To build a Hyperledger solution, you use the **Hyperledger Fabric** ledger software.

- With **Hyperledger Indy**, you can establish blockchain-rooted digital identities.

- **Hyperledger Grid** accelerates your blockchain project by providing many best practices for supply chains such as coded data models and others.

- With **Hyperledger Caliper**, you can access the performance of your blockchain solution.

- **Hyperledger Cello** can be used as operations console for your blockchain solution.

Hyperledger blockchain has been deployed at many clients, and you can find case studies on the previously referenced Hyperledger website. However, in the next section, we show you another example.

## Tradelens Uses Hyperledger Blockchain

In this section, we introduce a Hyperledger blockchain solution which is called **Tradelens**[8]. Tradelens has been created by Maersk and IBM and has dozens of other companies from the logistics ecosystem as partners already. Several governments also participate in the solution since clearing freight at a border is a government-controlled process. The goal of the Tradelens solution is to streamline the data flow between all participants of the logistics supply chain such as logistics companies, government border control authorities, companies operating large ports or airport terminals, and so on. While the data exchange is streamlined, it still needs to be secure so that you have proof when, for example, a container switches from one custodian in the supply chain to another.

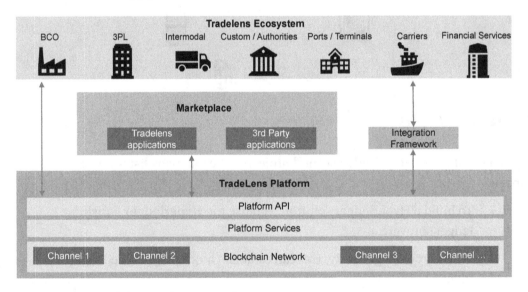

***Figure 11-4.*** *Tradelens Solution Architecture*

---

[8]More on Tradelens can be found here [10].

Another goal is to remove as much as possible of the paper-based, manual document processing. As a result, much better visibility of goods in the supply chain is achieved and freight in ports or terminals can be cleared faster with a higher degree of automation. Authorities benefit from complete insights into the transport chain with more accurate information and can focus their attention more on risk analysis and others.

The foundation of the Tradelens platform is Hyperledger blockchain where multiple channels can be created for different participants. On top of Hyperledger blockchain are the Tradelens platform and API services. Through the APIs, participants in the ecosystem of the Tradelens solution like custom authorities, beneficial cargo owners (BCO), ocean carrier companies, and so on can post transactions onto the blockchain when they received, for example, a container or cleared customs for freight. The solution architecture[9] is shown in Figure 11-4. With the new API-based access, a lot of the pain of the EDI framework which was previously used as the integration framework (gray box in Figure 11-4) can be increasingly avoided as more and more participants switch to the new, modern APIs.

Key components of the solution are running on the IBM Cloud. At the time of this writing, over 1 billion event transactions, more than 8 million documents, and over 21 million container transactions have been processed by this blockchain platform.

# On-Chain vs. Off-Chain Analytics

The main purpose of the blockchain technology is distributed transaction processing while establishing trust between the participants operating the peer nodes in the peer-to-peer network representing a blockchain. In the Hyperledger blockchain technology, the persistency is based on LevelDB or CouchDB, and the data is stored in binary or JSON format. A lot of analytics tools such as IBM Cognos, IBM SPSS, Tableau, and so on and data science tools like IBM Watson Studio or AWS SageMaker, which have been built over the last decades, have been optimized for data stores like relational databases supporting SQL. Many of these data stores have been optimized supporting advanced analytics like columnar databases. Another major approach for analytics is based on data lakes using the Hadoop platform with horizontal scale-out across a large number

---

[9]More details on the solution architecture, the data specification, and others can be found here [11].

of compute nodes. Thus, the architecture question is whether or not you can run AI effectively **on-chain** or **off-chain**. On-chain means you leave the data within the data repository of your blockchain system. Off-chain means you extract the data from the blockchain persistency and move it elsewhere where you can apply all your AI tools on it.

Let us first take a look at on-chain analytics. Running analytics on a peer node of a blockchain means that the analytics compete with the transaction processing of the peer for the same computational resources which are finite. As a consequence, a peer node will not be able to process the same transaction volume with the same performance if a portion of the resources are used by analytics compared to a scenario, where analytics are not consuming a portion of the resources. In almost any blockchain project, some basic reporting needs have to be satisfied like how many transactions were attempted, how many succeeded, how many failed, and so on. These are very basic and simple dashboarding needs, and all data required to answer these questions resides within the blockchain itself. Assuming your reporting tool of choice supports binary or JSON format and can integrate with the blockchain security features, such reports can be produced by reading the data directly on the peer. For example, the Hyperledger Explorer[10] gives you the capabilities to report on the number of peer nodes in the network, the number of successful transactions, and so on. The benefits of this approach are that the blockchain transactions never leave the blockchain peer network and hence remain secure. Also, you don't need to invest in data integration and messaging technologies.

However, the disadvantages of such an approach are also obvious: Many advanced AI capabilities which require significant computational resources can't be applied without significant performance and scalability impact to the transaction processing on the blockchain. Another reason is that the blockchain data is still in binary or JSON format and not transformed into a data format where analytics can be done with much better performance and scale. Also, many AI scenarios use data from a variety of sources, where blockchain transactional data is just another source. Therefore, if your AI solution requires data from multiple sources, on-chain analytics is architecturally not an appropriate solution because it is impractical to move the data from other sources onto the blockchain – a system designed for transaction processing and not for analytics. The only effective way to consume blockchain transaction data in a scenario where you want to keep the data on-chain and still be able to combine it with data from other sources is

---

[10]More details on Hyperledger Explorer can be found here [12].

through federation. IBM Db2 federation supports federated SQL queries across multiple different data sources, and in the most recent release, the support has been extended to peer nodes of Hyperledger blockchain.[11]

Now we take a look at off-chain analytics. Let us assume the following scenario from the insurance industry. Insurance companies and the insurance regulator collaborate using a blockchain network. The insurance companies place transactions on that blockchain network. Each transaction represents one claim which has been settled relative to natural disasters where the type of disaster and the amount paid are recorded. The natural disasters could be wildfires, floods, and so on. The insurance regulator can see all these transactions on the blockchain network and would be interested to predict under which circumstances of natural disasters one or more insurance companies need to file for bankruptcy because they can't settle the related claims anymore. For developing such prediction models, the insurance regulator needs to extract the transactions from the blockchain into an analytics environment where these transactions can be combined, for example, with live weather data from other sources, to determine if there are risks of large wildfires, floods, and so on. This is of course just one out of many examples where the blockchain transactions could be a relevant data source for AI.

Generally speaking, such scenarios require that the data is moved from a peer node into an analytics environment where the AI techniques are applied as shown in Figure 11-5.

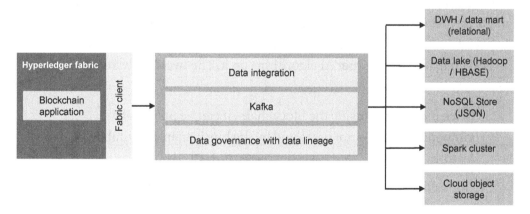

**Figure 11-5.** *Off-Chain Analytics*

---

[11]Details on how you can use these capabilities are found here [13].

Depending on the scenario, a participant running a peer in a blockchain network would like to move a portion or all of the data the participant is entitled to see to an analytics environment. As shown in Figure 11-5, there are many different possible targets such as data warehouses, data lakes, data marts, and so on. They can be located in private cloud or public cloud environments. The data transfer can be done as batch transfer using traditional data integration tools from vendors like Informatica or IBM or open source tools like Nifi or Airflow[12]. If you need the blockchain transactions as they are posted on the blockchain near real time replicated to your analytical systems, then you need messaging tools like the open source Kafka messaging system or commercial counterparts.

Hyperledger has an Event Hub mechanism where you can basically create subscriptions for transactions events which can be published onto the messaging bus and routed from there to your analytics system. Regardless if you do batch or ongoing data transfer or an initial batch with all further changes being continuously replicated, once the batch transfer finishes, the technology you deploy must support integration with the security architecture of your blockchain system.

For Hyperledger, this requires authentication and digital certificates integration. In off-chain analytics scenarios, you have also an additional advantage that you can transform the binary or JSON data from blockchain into formats much more suitable for AI. For example, you can use Kafka to seamlessly replicate from Hyperledger to IBM Cloud Object Storage on the public cloud as is to land the data next to other data assets you might have there. You can then use the SQL query[13] service to reformat the JSON blockchain data into a more analytics-friendly format and then run either large-scale SQL-based analytics or any other AI algorithms on it. The two disadvantages of this approach are

1. Once the data is outside the blockchain, it can be potentially tampered since it is not protected by the blockchain immutability features anymore.

2. Opposite to the on-chain analytics approach, you require data movement tooling, which comes at least with operational cost and potentially license costs if it is commercial software.

---

[12]More details on these two open source data integration tools can be found here [14] and here [15].

[13]You can try SQL query here [16].

In summary, the off-chain analytics approach supports a much broader set of use cases, has better scale and performance characteristics, and allows to execute substantially more AI techniques on the blockchain data. In the foreseeable future, this is very likely the most common approach to apply AI on blockchain data.

# Existing Technology Adopting Blockchain Concepts

While there are plenty of blockchain technologies available today, blockchain adoption is challenged for a number of reasons, and the following list is just a small subset of the concerns[14]:

- **Skills**: Introducing blockchain technology within an enterprise requires adoption of new skills by either training existing IT staff or hiring people with the right skills who are not easy to find.

- **Lack of industry standards**: In the area of transactional and analytical processing, SQL has been established as a standard and is supported by most vendors in the IT industry today. For blockchain technology, a similar standard and standard adoption is still outstanding.

- **Introduction of blockchain solution is disruptive**: Business processes and the IT landscape need to be reengineered to adopt a new business process which benefits from blockchain. This is disruptive to the existing IT infrastructure, and people familiar with the existing process need to be retrained on the new process and so on.

- **Limited transaction throughput**: The limited transaction throughput compared to (distributed) relational database management systems is also seen as a significant limitation of the blockchain technologies today. While substantial improvements occurred in recent years, there are still concerns if blockchain technologies ever achieve the transaction throughput required for widespread adoption.

---

[14]In this recent article from March 31, 2020, several more issues are discussed – see [17].

A technology which is not suffering from several of these concerns is (distributed) relational databases. There are plenty of skilled subject matter experts available, SQL has been standardized and is well supported, almost all data science tools support these sources remarkably well, and they excel at transaction processing with very high-throughput rates. Not surprisingly, vendors have taken first steps to combine the strengths of relational database and blockchain technologies. As a result, database vendors like Oracle[15] added blockchain capabilities to their existing relational database. Oracle introduced a capability they call *blockchain tables*. The key idea is to allow the use of append-only tables which applies reduction to create and read operations only against such tables with appropriate features to make them tamperproof and suitable for non-repudiation properties. The benefit of this approach is that companies can leverage blockchain-like capabilities in an environment which does not require cross-enterprise collaboration using infrastructure and skills they have already.

In addition to established commercial database vendors adding some blockchain capabilities to their existing database technology, a number of new vendors entered the blockchain database market such as BigchainDB, Fluree, ProvenDB, and AWS Quantum Ledger Database (QLDB)[16]. Common aspects are immutability, transparency, cryptography-based security, and consumability through SQL.

Internal fraud prevention, tamper protection against malicious DBAs or attacks where the DBA credentials got hacked and compliance use cases are just some example use cases where the blockchain-like capabilities added to the database provide significant value. Another advantage is that data being processed in database technologies, which have been enhanced with blockchain capabilities, have one significant advantage: a lot of data integration tools and AI tools support them extremely well. Blockchain capabilities in databases address certain blockchain use cases, but certainly not all of them, in particular use cases involving multiple enterprises. However, if your use case would allow to just use a database with blockchain features, your data science team will certainly be able to leverage data from such a source more easily for AI.

---

[15]More details on how the blockchain tables in Oracle databases work can be found in [19].
[16]More details on all these technologies are available here [20], [21], [22], and [23].

# Using Blockchain for AI Governance

As outlined in Chapter 8, "*AI and Governance*," AI governance is a growing concern. The European Union released an AI regulation[17] in February 2020 to address the following key questions:

- How can we ensure that data sets are not biased and sufficiently representative?

- How can we provide mandatory documentation of training, testing, data selection, and AI algorithms being used?

- How can we provide information to consumers about AI capabilities applied and their limitations in an automatic fashion?

- How can we ensure that AI decisions are robust, accurate, and reproducible?

- How can we integrate human oversight through approvals and monitoring into our AI-infused business processes?

To better understand why these questions are critical, think about the following: if AI-based prediction models are used by banks to determine if and under which conditions you might or might not get a mortgage or if insurances use AI-based models to determine whether or not your claim is accepted and how much money will be paid to you, you might want to know how reliable these models are, who trained them, and so on. Individual agents in robotic swarm ecosystems work toward a swarm goal. However, their decision making is decentralized and the decisions can be based on AI algorithms. The robots in such a use case make decisions and outcomes are based on majority rules. The vote based on an AI algorithm is basically the transaction. Again, how do you know how and why a conclusion was reached in a robotic swarm ecosystem using AI-based decision making by the robots? As outlined in Chapter 8, "*AI and Governance*," tools like IBM Watson OpenScale allow you to address some of the preceding questions like whether or not the training set was biased or if the model drifted over time and so on.

---

[17]You can find the regulation here [24].

IBM Watson OpenScale and similar tools do not address the need to provide a tamperproof audit trail with full transparency, who trained the machine learning model, which data sets was the model trained with, which version was used at a particular point in time, and so on. This is where AI and blockchain technologies nicely complement each other as shown in Figure 11-6.

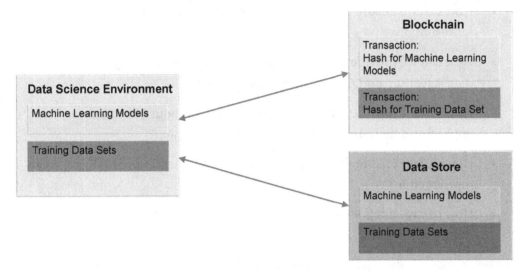

***Figure 11-6.*** *Blockchain Manages AI Models and Related Training Data Sets*

The key idea is basically to establish proof of existence on the blockchain, which ML model has been trained by whom and when on which data set, which version was deployed, and so on. Pushing large data volumes onto the blockchain which might have been used in your AI projects to train the models might not be possible. This can be either due to the volume or the size of individual items in structured or unstructured data sets. Hence, a more practical approach is to store the AI models and the training data sets in a data store outside of blockchain as shown in Figure 11-6. However, you push cryptographically secure hash values – conceptually a unique fingerprint of the model and the training data set alongside critical other attributes like who trained the model, version number of the model, and so on onto the blockchain itself. If you want to verify later that it was indeed a particular ML model and the training data set you have in the data store which was used, you could easily detect if someone tampered that because the hash value would not be the same as the hash value you find on the blockchain. So the immutability of the blockchain guarantees that you have a tamperproof way to show which AI model was running at the time a critical business decision based on AI-infused business processes was taken.

Going one step further, usually the input data for an ML-based prediction is relatively small. You can use the blockchain to create a transaction each time an AI model is used for a decision. Such a transaction on the blockchain would just need to include two things:

- The hash value of the used ML model, because all details about this model can be found by looking at the transaction where the hash of this model was originally recorded on the blockchain and optionally going from there to the external data store.

- The input values if they are small enough or a hash value of the input values. If the input values are too large, they could go to a data store outside of the blockchain where all the full values are recorded.

This approach, for example, would be useful for the robotic swarm ecosystem (decentralized system like the blockchain itself) or the AI-infused business process examples mentioned earlier. Blockchains can be characterized by the following traits: immutability, data integrity, resiliency to attacks, being deterministic, and decentralized. If you compare this to AI which is based on confidence in a prediction with a certain probability, volatility, continuous change (model needs to be retrained to avoid model thrift), and central decisions if you have the right model based on deep analysis, you can see that these two technologies are very different, but complementary to each other: if you bring them together though for AI governance, you achieve

- Enhanced data security

- Decentralized intelligence

- High efficiency

- Improved trust in AI-infused business process decisions or robotic swarm decisions

In a nutshell, using blockchains, you can improve your AI governance capabilities by creating a ledger of all AI-related artifacts and business decisions where AI was involved.

# Key Takeaways

We conclude this chapter as usual with several key takeaways. Blockchains produce transaction data on which AI should be applied for uncovering useful insights. This can be done to a more limited extent on-chain and to the full extent off-chain. We also showed that blockchain capabilities have been added to traditional technologies like databases offering for certain use cases within your enterprise viable alternatives compared to running full-blown blockchain solutions. Lastly, blockchain helps with establishing trust if you use them to establish proof on how your AI models have been developed and which version was used at a particular point in time.

*Table 11-1.*  *Key Takeaways*

| # | Key Takeaway | High-Level Description |
|---|---|---|
| 1 | Permissioned blockchains are for enterprise use cases | Permissioned blockchains like Hyperledger are private blockchains, where the identity of the participants is known. The consensus mechanism is substantially less resource intensive compared to cryptocurrency blockchains. Transactions are immutable |
| 2 | Blockchain applications produce transactional data | Blockchain applications are essentially transaction systems recording transactions related to physical or digital assets. Similar to any other transactional data, AI can be used to detect deep insights hidden in these transactions |
| 3 | On-chain vs. off-chain analytics | On-chain analytics compete with the transactions for the computational resources of the peer, and the persistency of blockchain technology is less optimized and suitable compared to other analytical environments. For the near- to mid-term foreseeable future, off-chain analytics are the better way to go because you can combine the blockchain transactions with other data sources as well as benefit from being able to apply all your AI capabilities in environments designed for scalable, high-performance AI |

*(continued)*

**Table 11-1.**  (*continued*)

| # | Key Takeaway | High-Level Description |
|---|---|---|
| 4 | Blockchain capabilities have been added to existing technologies | Unless you have blockchain use cases requiring collaboration across enterprises, you can leverage blockchain capabilities in database technologies where you have skills and existing infrastructure in house, providing lower-cost solutions |
| 5 | Blockchains can be used to provide trusted provenance of your AI-powered applications | The immutability feature of blockchains can be used for AI governance where you need to prove who, where, and when AI models have been trained and on which data sets this has been done |

# References

[1]   Satoshi Nakamoto: *Bitcoin: A Peer-to-Peer Electronic Cash System.*
      `https://bitcoin.org/bitcoin.pdf` (accessed April 2020).

[2]   Stuart Haber, W. Scott Scornetta: *How to Time-stamp a Digital
      Document.* `www.anf.es/pdf/Haber_Stornetta.pdf` (accessed
      April 2020).

[3]   Bitcoin: `https://bitcoin.org/en/` (accessed April 2020).

[4]   Ethereum: `https://ethereum.org/en/` (accessed April 2020).

[5]   Ripple: `https://ripple.com` (accessed April 2020).

[6]   Aleksander Berentsen, Fabian Schär: *A Short Introduction to the
      World of Cryptocurrencies.* `https://cdn.crowdfundinsider.com/
      wp-content/uploads/2018/01/St.-Louis-Federal-Reserve-a-
      short-introduction-to-the-world-of-cryptocurrencies.pdf`
      (accessed April 2020).

[7]   Cambridge Bitcoin Electricity Consumption Index: `www.cbeci.org`
      (accessed April 2020).

[8]   Hyperledger: `www.hyperledger.org/` (accessed April 2020).

[9]   Ala Al-Fuqaha, Nishara Nizamuddin, M. Habib Ur Rehman, Khaled Salah: Blockchain for AI: Review and Open Research Challenges. IEEE, 1.1.2019, p. 10127–10149, Electronic ISSN: 2169-3536, DOI: https://doi.org/10.1109/ACCESS.2018.2890507.

[10]   Tradelens: www.tradelens.com (accessed April 2020).

[11]   Tradelens solution architecture: https://docs.tradelens.com/learn/solution_architecture/ (accessed April 2020).

[12]   Hyperledger Explorer: www.hyperledger.org/projects/explorer (accessed April 2020).

[13]   Vinayak Agrawal, Sanjeev Ghimire: *Perform analytics on blockchain transactions.* https://developer.ibm.com/patterns/use-db2-and-sql-to-perform-analytics-on-blockchain-transactions/ (accessed April 2020).

[14]   Nifi: https://nifi.apache.org (accessed April 2020).

[15]   Airflow: https://airflow.apache.org (accessed April 2020).

[16]   *IBM SQL Query:* https://cloud.ibm.com/catalog/services/sql-query (accessed April 2020)

[17]   Boahua Yang: *10 Practical Issues of Blockchain Implementations.* www.hyperledger.org/blog/2020/03/31/title-10-practical-issues-for-blockchain-implementations (accessed April 2020).

[18]   Lukasz Golab, Christian Gorenflo, S. Keshav, Stephen Lee: *FastFabric: Scaling Hyperledger Fabric to 20,000 Transactions per Second.* https://arxiv.org/abs/1901.00910 (accessed April 2020).

[19]   Mark Rakhmilevich: *Blockchain Tables in Oracle Database.* https://blogs.oracle.com/blockchain/blockchain-tables-in-oracle-database:-technology-convergence (accessed April 2020).

[20]   BigChainDB: www.bigchaindb.com (accessed April 2020).

[21]   Fluree: https://flur.ee (accessed April 2020).

[22]   ProvenDB: `https://provendb.com/homepage/` (accessed April 2020).

[23]   Amazon Quantum Ledger Database: `https://aws.amazon.com/de/qldb/` (accessed April 2020).

[24]   European Commission: *On Artificial Intelligence – A European approach to excellence and trust.* `https://ec.europa.eu/info/sites/info/files/commission-white-paper-artificial-intelligence-feb2020_en.pdf` (accessed April 2020).

# CHAPTER 12

# AI and Quantum Computing

Richard P. Feynman, a Nobel Prize winner in physics, was a physicist thought leader in the areas of quantum mechanics and quantum electrodynamics. In 1982, he published a research paper with the title "Simulating Physics with Computers"[1]. In this paper, he asks the question if a quantum computer could be built (which he believed to be the case) or if classical computers can simulate the probabilistic behavior of a true quantum system (which he answered with a clear no). This research paper sparked interest in the scientific research community, which started to seriously explore whether or not a quantum computer can actually be built.

In this chapter, we explore the difference between existing processing architectures and a true quantum computer. We will then take a look at Shor's algorithm – a typical example of a problem where quantum computers excel – to help you understand which problems can be best tackled by quantum computers. Afterward we take a look at the status of AI and quantum computing today. We conclude the chapter with an outlook of how the field of quantum computing and AI is predicted to develop in the next several years.

## What Is a Quantum Computer?

There are many different processor architectures available today such as ASIC, FPGA, GPU, TPU, POWER, ARM, and so on as shown in Table 12-1. The optimization of these processor architectures followed Moore's Law for decades indicating that roughly every 18 months, the performance of a computer chip doubles.

---

[1]You can find the paper from Richard Feynman here [1].

© Eberhard Hechler, Martin Oberhofer, Thomas Schaeck 2020
E. Hechler et al., *Deploying AI in the Enterprise*, https://doi.org/10.1007/978-1-4842-6206-1_12

***Table 12-1.*** *Processor Architectures*

| Architecture | High-Level Description |
| --- | --- |
| Von Neumann | This is the general-purpose hardware architecture used in many desktop or laptop computers today. According to the Von Neumann rule, you cannot execute a fetch and data operation simultaneously in such a device which is performance-wise not optimal. The main benefit is that you can run a broad range of software on it – however not necessarily optimized for performance |
| ASIC | Application-specific integrated circuit (ASIC): these chips are optimized to embed the logic to execute software for a particular problem with an optimal hardware layout for best performance |
| FPGA | Field programmable gate array (FPGA): this is basically the middle ground between Von Neumann and ASIC architecture. It allows you to reconfigure the hardware (hence "field programable") to optimize performance to some extent, while the gate array portion of the name refers to the two-dimensional array of logic gates present in this architecture |
| GPU | Graphics processing unit (GPU):[2] opposite to CPUs, these chips are not optimized for latency, which means to finish a task as quickly as possible – but for throughput. Initially their primary application was to accelerate graphics for computer games. Today, they are also used in the space of high-performance computing (HPC) |
| TPU | Tensor processing unit (TPU): this architecture was originally developed by Google for accelerating Google's[3] machine learning framework known as TensorFlow |
| POWER | This processor architecture was originally created by IBM[4]. POWER is short for performance optimization with enhanced RISC and adopted also by others like Hitachi. Since 2019, the OpenPower Foundation[5] initiative is handled by the LinuxFoundation. IBM has a server business around the POWER architecture |

*(continued)*

[2]A quick introduction between CPU and GPU can be found here [2].

[3]An introduction to Google TPUs can be found here [3].

[4]See this reference [4] for further details.

[5]The homepage of the OpenPower Foundation can be found here [5].

***Table 12-1.*** (*continued*)

| Architecture | High-Level Description |
| --- | --- |
| ARM | This chip architecture is owned by ARM[6] Limited and the acronym ARM had different meanings over time (e.g., Advanced RISC Machines). The key design point of this architecture is low energy consumption while still delivering high performance. This makes it the ideal platform for embedded computing, and most of today's smart phones and tablet computers are running on ARM processors (iPhone, Android) |

However, this process is hitting the limits of physics now as the size of transistors has been reduced to a couple atoms scale and it can't be reduced further. Whether or not electrons flow through a transistor gate basically defines if the state is 0 or 1. Any further reduction in transistor size would allow the electrons to get to the other side of the transistor gate, while it is closed through a physical effect, called quantum tunneling which basically renders the transistor useless. While there are still increases in performance of computer systems possible via adding more cores, the process of decreasing the size of transistors is coming to an end.

Given all these different kinds of processor architectures available today, and the observation that there is not much promise in performance gains by making transistors smaller, you might wonder what's the point of creating yet another one which we call a quantum computer based on a quantum architecture?

To answer how the architecture of a quantum computer is fundamentally different, we need to dig deeper into the functionality of traditional computers. While the aforementioned processor architectures are different to some extent, they have a common base: they use bits to manage a state where a bit can hold a state of 0 or 1, where 0 on the physical level means that electricity is not flowing through the transistor gate and 1 means electricity is flowing. Specialized gates have been built on this basic principle, for example, a NOT gate gives you the inverse, the AND gate returns a 1 if and only if on both inputs there is an input signal, and so on. The counterpart of a bit in a quantum computer is known as a **qubit** (short for quantum bit).

---

[6]You can find more on the ARM company and the ARM architecture here [6].

In essence, a qubit is a complex number $c = a + bi$ that consists of a real number $a$ and an imaginary number $bi$, where $a^2 + b^2 = 1$. A qubit can store 2 traditional bits of information; 2 qubits can store 4 bits of traditional information; 3 qubits can store 8 bits = 1 byte of traditional information; with 30 qubits, you can store 128 MB; 31 qubits correspond to 256 MB; and so on. In other words, every qubit you add to a quantum computer doubles the amount of information it can process representing exponential growth. With approximately 45 qubits, you would be able to hold as much information as the largest super computers available today. Around 50 qubits you would reach quantum supremacy, the ability to compute problems not solvable by the largest supercomputers today.

Qubits have special properties known as **superposition** and **entanglement**. Let us take a look at them in the next two subsections.

# Superposition

Contrary to a traditional bit, a qubit holds combinations of 0 and 1 simultaneously based on the effect known as superposition. This means a qubit can hold multiple states at the same time – any proportion of 0 and 1. Note that a qubit does not have to be in any of the two states 0 or 1. Prior to a measurement, a qubit (represented by a complex number $a + bi$) can have any value on a circle with the radius of 1, as depicted in Figure 12-1 via the two dashed arrows. This qubit value can be deliberately changed. At the time of measurement, a qubit needs to decide if its value is 0 or 1 (that's the similarity to a traditional bit). This decision can, for instance, be done in a very straightforward way as follows: if $a^2 > 0$, the qubit value is 1 with certainty, and if $a = 0$, the qubit value is 0 with certainty.

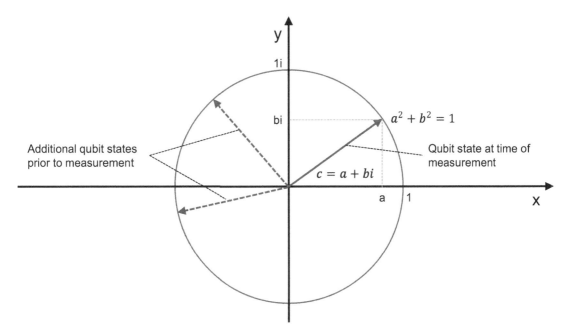

**Figure 12-1.** *Qubit as a Complex Number*

In quantum computers, whether or not the qubit is in state 0 or 1 when measured depends on quantum probabilities. The quantum probability is something very different than what many people typically consider based on probabilities experienced in life.

For example, if you throw a perfect coin, there is a 50% probability that the coin lands on either side. In addition, the probability of each result adds up to 1. If you roll a 6-sided dice, each number has a 1/6 chance to end up on top, and if you add the results, they again add up to 1. In such scenarios, there is a particular condition at play. If we assume you have probability $a$ and probability $b$, then the following conditions always holds true: $a + b \geq a$. This is based on the fact that the probability calculations in these examples are only using real numbers. In the world of quantum computing and qubits, probability is calculated based on complex numbers. As mentioned, a complex number is given by $c = a + bi$, where $a$ and $b$ are real numbers and $i$ is the imaginary unit. To understand the probability aspect during measurement of a qubit value, consider the example of Figure 12-2 showing the double slit experiment.

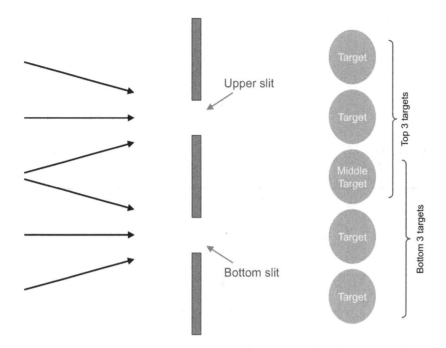

***Figure 12-2.*** *Double Slit Experiment*

First, we use the setup of Figure 12-2 for a traditional consideration. Assume we have a perfect shooter never missing the target. There is a 50% probability that the shooter shoots through either one of these slits. The double slit experiment setup is such that through the upper slit, the shooter can hit the three targets from the top, and through the slit at the bottom, the shooter can hit the three targets from the bottom. This means the target in the middle can be hit through both slits. If we calculate the probability that any one target is hit, we get the probability of 1/6 for each of the 2 targets from the top and the bottom and a 1/3 (1/6 + 1/6) for the target in the middle.

Now let us consider the same setup, but not using bullets complying with classical laws in physics. Instead, we now use a laser firing photons, which comply with the laws of quantum mechanics. We do the same calculation again using complex numbers. In this case, we want to know the probability of the photons hitting the target in the middle. If you calculate this, which requires to use complex numbers, you will find that the probability of a photon hitting the target in the middle is zero. This is the required calculation[7] for adding the probabilities hitting the target in the middle:

$$\frac{1}{\sqrt{2}}\left(\frac{-1+i}{\sqrt{6}}\right) + \frac{1}{\sqrt{2}}\left(\frac{-1-i}{\sqrt{6}}\right) = 0$$

The phenomenon behind this is called interference. At the beginning of this section, we stated that a qubit while in superposition has all possible states representing a proportion of 0 and 1 at the same time. Here, actually our photon is in many positions at once – the superposition. It's going through both slits at the same time. On the other side, it can cancel itself out. This is the interference. If you throw stones into water, you can observe that sometimes the waves are adding up and sometimes – due to interference – cancel each other out. Interference is the reason why the probability calculation in the quantum world does not comply to $a + b \geq a$ for two given probabilities $a$ and $b$, since they are now complex numbers.

In a classical computer, if you have 8 bytes, you can store *one* out of 256 states at any given point in time. With a quantum computer with 8 qubits, you can store *all* 256 states simultaneously. In other words, with superposition, quantum computers can hold many states at once. Imagine if you put a quantum computer into many different states of a classical computer, but at a given point in time. If you can then execute operations with all these states simultaneously, you have the ultimate parallel processing, which is the biggest promise of quantum computing.

## Entanglement

Two or more quantum particles can get entangled with each other. Lasers have been used to create quantum particles in an entangled state. Once quantum particles are entangled, they are perfectly linked. If one quantum particle in a set of entangled quantum particles changes its state, the other instantly does too. The interesting aspect is that this holds true across arbitrary large distances. Applications of entanglement can be found in quantum communication and quantum cryptography. In quantum communication, the idea is

---

[7]If you are really interested to get the basics of calculating this, we recommend you to read Noson Yanofsky's "Introduction to Quantum Computing." The preceding formula is from Noson Yanofsky's excellent introduction on this subject. With just a tiny bit of basic graph concepts and matrix calculations, this paper introduces the foundation of quantum computing including one of the most basic quantum algorithms, the Deutsch algorithm. Using the double slit experiment for illustrating interesting angles of superposition was inspired by this article. You can find the paper here [7].

to use entanglement between two communicating parties so that the receiver can only process the transferred qubit if sender and receiver have been put into an entangled state before. If someone in the middle would try to eavesdrop into the communication, the eavesdropper can't do anything with the intercepted qubit because the eavesdropper lacks the second qubit to decode it. Any attempt to understand the qubit by taking a measurement would alert sender and receiver of the presence of the eavesdropper since this would cause the collapse of the entangled state.

## Quantum Computer

With the understanding of superposition and entanglement, let us now define what we mean with a quantum computer: a quantum computer is a computer system based on quantum mechanical effects. It uses qubits with superposition and entanglement characteristics to be able to represent multiple states simultaneously. At the time of a measurement with a certain quantum-based probability, the read operation will yield a 0 or 1 measurement. Due to its quantum mechanical core, this is truly a new class of compute system.

To actually build and realize a qubit, there are different physical approaches[8] pursued today such as

- Photon-based quantum computing systems are building qubits using techniques where laser light beams are polarized. An advantage of this approach is that photons are only very weakly interacting with their surroundings making qubits based on this approach more stable.

- Ion-trapped quantum computing systems use ions or charged atomic particles suspended in electromagnetic fields to realize qubits. These systems require cooling to very low temperatures, oftentimes close to 0 Kelvin.

Figure 12-3 shows a quantum computer. The quantum compute chip is at the lowest part of the shown device. Warm air as we all know is moving up, so having the quantum chip in the lowest part of the device shown is having it in the coolest spot.

---

[8]If you want to get a summary overview, this research paper would be a good starting point [8].

***Figure 12-3.*** *A Quantum Computer Shown on IBM Think Conference*[9]

# Shor's Algorithm

To illustrate the potential of a quantum computer and its unique capabilities brought to computation with the effects of superposition of qubits, we want to introduce you to Shor's algorithm. The algorithm of Peter W. Shor[10] is a polynomial time algorithm to compute the prime factors of a given integer number on a quantum computer. The implications of the previous sentence potentially affect a lot of people around the world. To understand this, we provide in a first step a rather short introduction to the mathematical background of the problem. In a second step, we show where this mathematics is heavily used today to illustrate the tremendous impact of Shor's discovery.

---

[9]Picture taken by one of the authors at the IBM Think Conference in February 2019.
[10]You can find the research paper from Peter W. Shor here [9].

A number is considered to be a prime number if it can be divided only by 1 and the number itself without rest. For example, 7 is a prime number. It can only be divided by 1 and itself without rest. And 14 is not a prime number because 2 * 7 = 14, which means it can be divided by other numbers without rest. In this example, 2 and 7 are both prime numbers and represent the prime factors of 14. So the factorization of 14 into its prime factors is comprised of 2 and 7. While this looks like a simple task for 14, now take a look at the number 2606663. Its prime factors are 1249 and 2087. If you would need to try all the various combinations of prime numbers to find out which pair yields 2606663, you get a sense that there are many possibilities to test because you have many different combinations to explore.

A well-known theorem of number theory states that for any positive natural number, a factorization exists which is unique except for the order of the prime factors which does not matter (this means in the previous example, it does not matter if you write 14 = 2 * 7 or 14 = 7 * 2). This theorem is known by many different names like fundamental theorem of arithmetic or unique factorization theorem. The essential parts were known to Euclid, a mathematician from Greece who lived in the third century before Christ who published it in his famous *Elements*. Carl Friedrich Gauss, a German mathematician from the eighteenth century, published a more complete version in his famous *Disquisitiones Arithmeticae*.

Computer scientists in a field called complexity theory[11] classify algorithms in different categories of complexity considering how much time and space it takes to solve the hardest instance of a particular problem. Now, it can be shown that testing if a given number is a prime number can be done efficiently, in terms of complexity theory, this implies in polynomial time, which is the category of problems with lowest time and space complexity.

So let us assume you have two very large prime numbers $p_1$ and $p_2$. Consider now a number $q = p_1 * p_2$. Let us now assume you have only $q$ and you want to know what is the factorization of $q$; in other words, you want to compute $p_1$ and $p_2$, using a factorization algorithm. On classical computers, there is no known factorization algorithm which is able to compute this efficiently, meaning in polynomial runtime. None of the known

---

[11]A good introduction to this topic can be found here [10].

factorization algorithms like number field sieve, quadratic sieve, elliptic curve method[12], and so on runs in polynomial time. This observation is the foundation for many of the relevant cryptographic security features. Popular encryption algorithms like RSA[13] used to protect your privacy (sensitive data, communication channels, etc.) are relying on the fact that factorization is computationally a difficult and time-consuming problem even on the largest supercomputers available today. If you pick sufficiently large prime numbers, it could take thousands of years or more to break your encryption key.

The algorithm of Peter Shor basically shows that using a quantum computer, the same problem of factorization is getting easy to solve, which essentially means in polynomial time. What Shor's Algorithm does is the following:

1. **Prime factorization**: Shor shows how to reduce the general problem of prime factorization to a factorization problem.

2. **Factorization**: He then shows how to reduce the factorization problem to the problem of determining the order of a number.

3. **Determine order of a number**: On classical computers, this is a problem for which no polynomial algorithm exists today.

4. **Calculate order of a number**: Based on a quantum computation in polynomial time, the order of a number can be calculated.

Since the reduction steps from 1 to 3 are polynomial in runtime and 4 is in polynomial runtime as well, Shor's overall algorithm is a polynomial algorithm where step 4 only works on quantum computers. When we introduced the qubit with the superposition trait, we stated that it can hold any proportion of 0 and 1 simultaneously which means a lot of different combinations. The factorization problem is hard on a traditional computer because there are so many different combinations of prime numbers which need to be tested.

---

[12]For topics on cryptography, *Applied Cryptography* by Bruce Schneier is a must-read book. It also contains a wonderful section on the factorization problem and encryption algorithms using it. For the methods mentioned here, see the references [11], [12], [13], and [14], respectively.

[13]This public key encryption algorithm is named after its inventors Ron Rivest, Adi Shamir, and Leonard Adleman.

A sweet spot for quantum computing are problems where you need to find solutions within a very large number of possibilities. Shor's algorithm basically exploits the computational speedup by being able to use superposition of the qubits. So if quantum computers get mature enough (a question we need to discuss later in this chapter), Shor's discovery basically obliterates any security which relies on encryption algorithms exploiting factorization to protect sensitive data and secure communication channels used by enterprises and by millions of people every day whether they are aware of it or not. When Shor published his research paper, it made headline news because of this threat. Experts in cryptography are working on quantum computing-resistant cryptography capabilities already today. A promising technique seems to be lattice cryptography[14]. The idea is to have a cryptographically secure method which allows us to secure our data even if quantum computers get into mainstream at some point in the future.

# AI and Quantum Computing Today

Many IT companies are heavily investing in quantum computing today, among them Google, Alibaba, Microsoft, and IBM. For instance, IBM, the German Government, and the Fraunhofer Research Institute agreed to bring quantum computing in a joint initiative to Germany for companies and researchers to explore the possibilities of quantum computing. On March 13, 2020,[15] the Fraunhofer Research Institute and IBM announced that IBM is installing a quantum computer in an IBM facility in Ehningen (close to Stuttgart) within Germany which will be made accessible through the IBM Cloud to companies and researchers. IBM calls its quantum computer IBM Q System One and released a first commercial 20-qubit-based version of it in 2019. Late in 2019, the next major version of IBM Q System One had 53 qubits already. Since 2016, IBM offers access to its System Q through the IBM public cloud and allows researchers and companies to explore the possibilities of quantum computers remotely already since then. To develop algorithms for the quantum computer, there is a rich set of tools available in an open source package called Qiskit[16]. The ecosystem around System Q seems to be strongest at this point with hundreds of thousands of downloads of Qiskit by researchers and developers around the world, exploring the possibilities of quantum computing with it.

---

[14]Matthew Dozer produced a nice video introducing lattice-based cryptography and how it can be used which you can find here [15].

[15]You can find the press announcement from the Fraunhofer Research Institute here [16].

[16]You can download Qiskit here [17].

Another leader in the race to quantum computing is Google. Google announced that their quantum compute team has proven quantum supremacy in a research paper[17] published in 2019. Quantum supremacy is a term which got popular through the physicist John Preskill in 2012 and refers to the point where a quantum computer performs a calculation significantly faster than any supercomputer today. On a quantum computer with 53 qubits named *Sycamore*, the quantum computer scientists from Google ran a calculation in only 200 seconds, which on the fastest supercomputers today would need 10000 years. Shortly after the release of this paper, a heated discussion started whether or not the computation would really need 10000 years. That's however beyond the point. In 2019, both Google and IBM, with quantum computers comprised of 50 qubits and more, are demonstrating that quantum computing is indeed holding a promise of being able to solve problems, which are not solvable with today's traditional computers. The interesting question from an AI perspective is what sort of problems are suited for quantum computing today. Table 12-2 provides a short summary overview of AI relevant use cases where quantum computing is believed to have a huge potential.

***Table 12-2.*** *AI Use Cases for Quantum Computing*

| Type of Algorithm | Use Cases | Example |
|---|---|---|
| Optimization | Logistic routing | Optimize flight schedules |
| | Operations optimization | Optimize deliver routes |
| | Supply chain optimization | Increased productivity by optimized resource usage |
| Scenario/what-if simulation | Market prediction | Asset value evaluation for trades |
| | Risk prediction | Economic impact of significant variable changes in economic systems |
| Molecular simulation | Molecular design | Engineer drugs or materials for particular use cases |
| | | Optimize existing chemicals, for example, batteries for electric cars |

(*continued*)

---

[17]You can find the research paper from the Google scientists here [18].

***Table 12-2.*** (*continued*)

| Type of Algorithm | Use Cases | Example |
| --- | --- | --- |
| Machine learning/data mining | Pattern discovery Prediction Classification | Discover anomalies in data |

The example of logistic routing in Table 12-2 comes from the logistics industry. If you need to plan the scheduling of flights, trains, delivery trucks, and so on, you are dealing with an optimization problem which has many possibilities, and the difficulty lies in finding the optimal solution. For example, delta is exploring quantum computing to see where better results on optimization problems can be achieved. A key algorithm in this area is the *Quantum Approximate Optimization Algorithm.*[18]

Simulating the behavior of molecules which is another use case of Table 12-2 is a very tough and challenging problem. However, without precise simulation, new drugs and materials cannot be discovered. For the automotive industry, electric cars are considered a strategic element for the future. The essential component of an electric car is the battery. The amount of electricity it can store and the speed it can be recharged are critical to improve market acceptance of electric cars. Daimler collaborates with IBM using IBM quantum computers and in January 2020 found the first promising results[19] using molecule simulation on a quantum computer. For material science, the simulation of more complex molecule structures requires supercomputing infrastructure which consumes a lot of energy. One promise of quantum computing is that these simulations cannot only be done much faster on a quantum computer (keep in mind that the number of states a quantum computer can hold and parallelly operate on grows exponentially) but more importantly at a significant lower electricity bill. Simulations which can be expressed by the Hamilton Operator (also known as *Hamiltonian simulations*[20]) can be done efficiently by quantum computers, a result which was proven by Seth Lloyd in 1996. This type of simulation

---

[18]You can watch this YouTube video to get an introduction on this topic [19].

[19]You can read more on this here [20].

[20]Watch this YouTube video to get an introduction on the topic [21].

is what traditional supercomputers cannot do efficiently. This type of AI application is hence one area where quantum computing holds a great future promise. The Variational Quantum Eigensolver[21] discovered in 2014 is another type of simulation algorithm which runs extremely well on quantum computers.

For supervised DL and DL techniques, which we have elaborated on in this book, quantum computing researchers are working on neural quantum nets.[22] In 2018, Edward Farhi and Hartmud Neven[23] showed as first results how these AI techniques could be used on quantum computers using neural quantum nets.

Another important use case of quantum computing is the quantum Internet. A quantum Internet is an Internet of connected quantum computers. Communication lines between the quantum computers can use telecommunication fibers. However, standard optical switches cannot be used since they need to preserve the state of the qubits being transferred to avoid decoherence which negatively affects the quantum state of a qubit potentially rendering it useless. For transfers over long distances, quantum repeaters are needed since normal repeaters will not work. A key element used in a quantum Internet is the entanglement between qubits. What might sound like science fiction is already explored today. In the Netherlands, KPN and QuTech agreed in 2019[24] to build a quantum Internet connecting several cities in the Netherlands. Similar efforts are already underway in Japan, China, and Switzerland since several years. Among the many potential benefits of a quantum Internet, the most outstanding one is that it is truly secure. As mentioned in the entanglement section earlier, if someone tries to eavesdrop into a quantum communication, this is not going undetected and any attempt to read the intercepted qubit renders it useless.

Despite all these interesting use cases and the substantial amount of progress in quantum computing over the last decades, quantum computing is not a mainstream technology today. It still needs to overcome a couple of very substantial problems. The first one has to do with an effect called **quantum decoherence**. Basically, the qubits used in quantum computers are very sensitive to influences from their surrounding environment and must be shielded as much as possible. However, it cannot be perfectly shielded. Due to that,

---

[21]This paper provides you more details on Variational Quantum Eigensolver algorithm [22].

[22]Nielsen wrote two books on deep learning and quantum computing which are excellent introductions on both topics [23] and [24]. A good introduction on quantum machine learning is [25].

[23]You can find their paper here [26].

[24]The announcement from KPN and QuTech is located here [27].

the state of qubits deteriorates over time. To perform manipulations on qubits (after all, you want to calculate with them) or to measure the result of a quantum computation access is required, and hence the shielding can never be perfect.

If you need multiple qubits for your quantum calculation, this problem of quantum decoherence becomes even more severe. Figure 12-4 essentially says that current designers of quantum computers are struggling with the following challenge: the more qubits they involve in their quantum computation, the shorter the time they can use them for performing a quantum calculation. This means you have to make a choice: you can process larger volumes of data, but then you can only work on it for a shorter amount of time. The alternative is to use a smaller amount of data with increased calculation time.

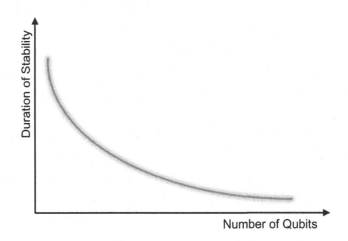

***Figure 12-4.*** *Problem of Quantum Computing*

On classical computers to deal with errors by processing with bits, the state of a bit can be copied to another bit. By the *no cloning theorem* found in 1982, the same approach unfortunately is not available to quantum computers. Researchers must thus find different approaches for quantum computing to handle quantum error correction. One promising idea is to use multiple qubits as a single logical qubit. If one of the qubits fails on the physical level, this qubit can get corrected by other qubits belonging to the same logical qubit.

Another problem of quantum computing is that whenever you perform a measurement to obtain the result, the state of the qubits gets destroyed and you need to initialize again.

For big data type use cases, there is also the challenge on how you load large data volumes into a quantum computer and more specifically into a quantum random access memory (QRAM[25]). The design and effective implementation of QRAM structures is unfortunately also not a trivial task and still presents a challenge for the hardware design aspects of quantum computers.

As mentioned earlier in this chapter, certain types of quantum computers also require cooling to temperatures just slightly above 0 Kelvin. Cooling infrastructure to such low temperatures is nothing you can install in an office or at home today.

As a result, despite all its progress over the last decades, quantum computing is still in its infancy today and not ready for mainstream consumption and enterprise deployments.

# AI and Quantum Computing Tomorrow

To conclude this chapter, we want to give you a perspective on the outlook of AI and quantum computing.[26] If you search online, you will find that there are some voices arguing that the hurdles to build quantum computers for a broader commercial exploitation are a futile attempt. While these negative voices do exist, there are also a large number of scientific researchers who believe there is some promising and substantial progress possible. In this section, we will focus on the more positive outlook. Depending on the source, mainstream and enterprise adoption for quantum computers is predicted to be 5–10 years away as shown in Figure 12-5.

At this stage, as you can see in Figure 12-5, quantum technology is in a phase of early exploration. Right now, quantum computers are manufactured in low quantities. For example, in the announcement from the Fraunhofer Institute and IBM as referenced earlier, it was mentioned that IBM built 15 quantum computers with 20 or more qubits to date, which are accessible through the IBM public cloud for companies and researchers to explore quantum computing.

On the hardware side, many severe challenges need to be solved in the next couple of years before quantum computing can be transitioned to mainstream technology. For example, to increase the number of qubits in a quantum computer further, the problem of decoherence needs to be mitigated so that computations can run for extended time periods.

---

[25]More details on QRAM can be found in this paper [28].
[26]In this YouTube video, you can find the perspective of Seth Lloyd [29].

On the software side, software needs to be created which can be used as an efficient platform to develop software for quantum computers. From a software perspective, we need enhancements across the full stack: this covers operating systems and system management software components as well as runtime environments on which the researcher and developer community can effectively build commercial software solutions.[27]

Due to these challenges, it is predicted that only around 2030 the transition of quantum computing becomes ubiquitous.

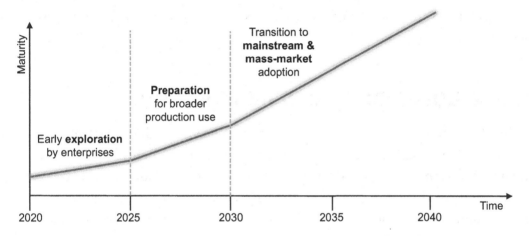

***Figure 12-5.*** *Outlook of Quantum Computing Becoming Mainstream Technology*

While quantum computing has its biggest promise to provide significant performance improvements for combinatorial AI problems, it is currently not yet clear if quantum computers will replace all traditional compute systems. Similar to a desktop or server computer today, which have CPUs as well as other chips like GPU for acceleration of graphic processing, or crypto co-processors for hardware accelerated crypto functionality, the future of quantum computing might be in hybrid hardware architectures as shown in Figure 12-6.

---

[27]A very good overview on necessary hardware and software enhancements for quantum computing can be found in this research article from 2019 [30].

A quantum chip would be just another accelerator for certain workloads in a broader system architecture providing specialized acceleration for certain types of AI use cases. For such hybrid architectures, problems which still need to be solved are communication and data transfer between the new quantum chip and the other parts of the system.

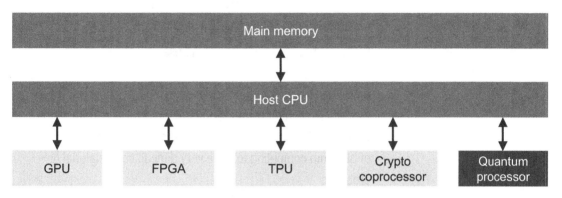

***Figure 12-6.*** *Hybrid System Architecture with Quantum Processors*

While it is definitely too early to say for sure whether or not quantum computing makes it to mass market adoption, for AI applications, quantum computing as a novel compute platform holds a yet to be explored promising future.

# Key Takeaways

We conclude this chapter as usual with several key takeaways, summarized in Table 12-3. As you have seen, a quantum chip using qubits is an entirely new computing paradigm and not just yet another chip architecture. For certain types of combinatorial AI problems, quantum computing holds the promise to be substantially faster compared to all traditional computing systems known today. At this point, quantum computing is in its early stages with mass market readiness likely away 5–10 years.

***Table 12-3.*** *Key Takeaways*

| # | Key Takeaway | High-Level Description |
|---|---|---|
| 1 | Qubit, superposition and entanglement | A qubit for a quantum computer is the quantum mechanical equivalent to a bit in a traditional computer system. The special powers of a quantum computer are the characteristics of superposition and entanglement of qubits |
| 2 | Algorithm of Shor | Peter W. Shor showed that with the use of quantum computers, the factorization problem can be efficiently solved in polynomial time. This is a major threat for all encryption systems relying on the factorization problem to be computationally very difficult on traditional computer systems. This example illustrates the power of quantum computing to solve very difficult combinatorial AI problems |
| 3 | Some main use cases for quantum computing are molecular simulation and optimization | Combinatorial AI problems are the sweet spot for quantum computing. These problems are found in many industries like pharmaceutical, automotive, banking, insurance, logistics, and so on |
| 4 | Quantum computing is in its early stage | Due to problems like decoherence, among others, it is predicted that quantum computing is 5–10 years away from mass market readiness and enterprise exploitation. That being said, you should explore quantum computing today to identify which use cases it promises significant value to your enterprise |

# References

[1] Richard P. Feynman: *Simulating Physics with Computers.* International Journal of Theoretical Physics, 21, p. 467–488, 1982.

[2] New York University: *Graphical Processing Unit (GPU) Introduction.* https://nyu-cds.github.io/python-gpu/01-introduction/ (accessed April 2020).

[3] Google Tensor Processing Unit: https://cloud.google.com/tpu/docs/system-architecture (accessed April 2020).

[4]   IBM Power systems: www.ibm.com/it-infrastructure/power/power9 (accessed April 2020).

[5]   OpenPower Foundation: https://openpowerfoundation.org (accessed April 2020)

[6]   ARM company: www.arm.com/why-arm/architecture (accessed April 2020).

[7]   Noson S. Yanofski: *An Introduction to Quantum Computing.* https://arxiv.org/abs/0708.0261 (accessed April 2020).

[8]   Hamid Reza Bolhasani, Farid Kheiri, Amir Masoud Rahmani: *An Introduction to Quantum Computers Architecture.* www.researchgate.net/publication/337144719_An_Introduction_to_Quantum_Computers_Architecture (accessed April 2020).

[9]   Peter W. Shor: *Polynomial-Time Algorithms for Prime Factorization and Discrete Logarithms on a Quantum Computer.* In: SIAM Journal on Computing, 26/1997, p. 1484–1509.

[10]  D. P. Bovet and P. Crescenzi: *Introduction to the Theory of Complexity.* Englewood Cliffs, N.J.: Prentice Hall, 1994.

[11]  Bruce Schneier: *Applied Cryptography.* 2nd Edition, John Wiley and Sons, 1996.

[12]  A. K. Lenstra and H.W. Lenstra Jr. eds.: *Lecture Notes in Mathematics 1554: The Development of the Number Field Sieve.* Springer Verlag, 1993.

[13]  C. Pomerance: *The Quadratic Sieve Factoring Algorithm.* Advances in Cryptology: Proceedings of EUROCRYPT 84, Springer Verlag, 1985, p. 169–182.

[14]  *H.W. Lenstra Jr.: Elliptic Curves and Number Theoretic Algorithms.* Report 86-19, Mathematisch Instituut, Universiteit of Amsterdam, 1986.

[15]  Matthew Dozer: *Introduction to lattice-based cryptography.* www.youtube.com/watch?v=37Ri1jpl5p8 (accessed April 2020).

[16]   Fraunhofer Institut: `www.fraunhofer.de/de/presse/`
`presseinformationen/2020/maerz/fraunhofer-und-ibm-`
`bringen-quantenrechner-fuer-industrie-und-forschung-`
`nach-deutschland.html` (accessed April 2020).

[17]   Qiskit: `https://qiskit.org` (accessed April 2020).

[18]   Frank Arute, Kunal Arya, Ryan Babbush, Dave Bacon, Joseph
C. Bardin, John M. Martinis et al.: *Quantum supremacy using a*
*programmable superconducting processor.* Nature, Volume 574,
October 2019, p. 505–510, doi:https://doi.org/10.1038/s41586-
019-1666-5/

[19]   Edward Farhi: *A Quantum Approximate Optimization Algorithm.*
`www.youtube.com/watch?v=J8yoVhnISi8` (accessed April 2020).

[20]   Jeanette Garcia: *IBM and Daimler use quantum computer*
*to develop next-gen batteries.* `www.ibm.com/blogs/`
`research/2020/01/next-gen-lithium-sulfur-batteries/`
(accessed April 2020).

[21]   Isaac Chuang, Guang-Hao Low: *Optimal Hamiltonian*
*Simulation by Quantum Signal Processing.* `www.youtube.com/`
`watch?v=Cv9juBFHIVs` (accessed April 2020).

[22]   Ryan Babbush, Alan Aspuru-Guzik Jarrod McClean, Jonathan
Romero: *The theory of variational hybrid quantum-classical*
*algorithms.* `https://arxiv.org/abs/1509.04279` (accessed
April 2020).

[23]   Michael Nielsen: *Neural networks and deep learning.*
`http://neuralnetworksanddeeplearning.com` (accessed
April 2020).

[24]   Isaac Chuang, Michael Nielsen: *Quantum Computing and*
*Quantum Information.* Cambridge Series on Information and
the Natural Sciences. Cambridge University Press, ISBN-13: 978-
0521635035, 2000.

[25]   F. Petruccione, M. Schuld, I. Sinayskiy: *An introduction to quantum machine learning.* `https://arxiv.org/abs/1409.3097` (accessed April 2020).

[26]   Edward Farhi, Hartmut Neven: *Classification with Quantum Neural Networks on Near Term Processors.* `https://arxiv.org/abs/1802.06002` (accessed April 2020).

[27]   KPN and QuTech join forces to make quantum internet a reality. `www.overons.kpn/en/news/2019/kpn-and-qutech-join-forces-to-make-quantum-internet-a-reality` (accessed April 2020).

[28]   Vittorio Giovannetti, Seth Lloyd, Lorenzo Maccone: *Quantum random access memory.* `https://arxiv.org/abs/0708.1879` (accessed April 2020).

[29]   I. Ashraf: K. Bertels, T. Hubregsten, A.Krol, A.A. Mouedenne, A. Sarkar, A. Yadav: *Quantum Computer Architecture. Towards Full-Stack Quantum Accelerators.* `https://arxiv.org/pdf/1903.09575.pdf` (accessed April 2020).

[30]   Seth Lloyd: *The Future of Quantum Computing.* `www.youtube.com/watch?v=5xW49CzjhgI` (accessed April 2020).

# PART IV

# AI Limitations and Future Challenges

# CHAPTER 13

# Limitations of AI

The promise of AI with its breathtaking range of applications seems to be without limits. To elaborate on *limitations of AI* may therefore be perceived by some of our readers as a spin in opposite directions. AI is so much associated with accelerating innovation, insight, and decision making that we see its opportunities as immeasurable. And yet, even for AI, there are limits and challenges, as we learn about in this chapter.

A simple, straightforward business problem may require analytics and insight for adequate decision making; however, using AI with ML or DL models may be inappropriate, not meaningful, and adding an unnecessary degree of complexity without delivering new insight. In some cases, the reality and circumstances may be highly susceptible to change, and before an ML model can be learned and deployed, it may already be meaningless. As we have seen in Chapter 8, *"AI and Governance,"* with risk management and compliance, there may also be some legal reasons or compliancy regulations that may prevent or at least limit AI applications for specific scenarios.

## Introduction

In this chapter, we focus on current AI limitations and challenges that either prevent us from using AI (e.g., if multitask capability in generalized learning is required) or at least limit the AI applicability (e.g., because of too costly data labeling and annotation efforts). Autonomous learning, meaning self-directed learning, taking control of one's own learning behavior to adapt to new circumstances, and blending learning with preferences, opinions, views, and others, is another area where humans may want to stay in control.

© Eberhard Hechler, Martin Oberhofer, Thomas Schaeck 2020
E. Hechler et al., *Deploying AI in the Enterprise*, https://doi.org/10.1007/978-1-4842-6206-1_13

We also discuss insoluble challenges, such as missing understanding and reasoning of ML and DL models. Our way of dealing with *limitations of AI* is linked to applications and scenarios, where we are hitting current limitations and insoluble challenges by AI and DL. Especially the insoluble challenges may truly lead us not to use AI – even in the far distant future. In this context, we also share with you some AI research topics, such as *learning to learn* (meta-learning), which will eventually improve the AI applicability scope.

AI limitations are strongly related to understanding and comparing AI and DL with the human brain,[1] particularly in the context of its cognitive and learning capabilities. In many cases, a human being is of the essence and may never be entirely replaced by AI – even not in the far distant future. In our discussion on *limitations of AI*, we intend to make you sensitive with regard to *how to use AI* properly and when to use *which* AI capabilities and when better not.

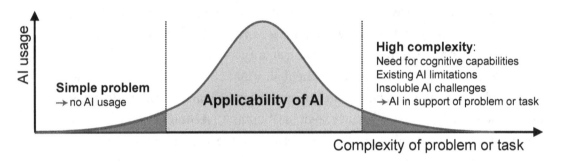

**Figure 13-1.** *When Not to Use AI*

As illustrated in Figure 13-1, to understand *limitations of AI* is discussed in connection with the exceptional and unsurpassed capabilities of the human brain (in particular the cognitive capabilities), current technical gaps or deficiencies of AI, and also insoluble AI challenges. There are numerous scenarios, which require the unique abilities of a human being with its cognitive capabilities and generalized learning skills or which strike AI limitations, such as entrepreneurial intuition, creativity and novel innovative approaches, understanding and interpreting decisions, and reasoning of situations or circumstances, to just name a few.

---

[1]See [1] for more information on comparing the human brain with a computer and AI.

The applicability of AI is large, as you see illustrated in the middle area of Figure 13-1. Toward the left are relatively simple problems that may clearly not benefit from AI usage, whereas toward the right side, you see high complexity problems and tasks that may require, for instance, cognitive capabilities or where current AI limitations and insoluble AI challenges may either prevent adequate AI usage or would allow AI to be "only" in support of a human being.

# AI and the Human Brain

Understanding the current limitations of AI and some insoluble challenges for AI suggests to perform a basic comparison of the human brain with an artificial neural network, which is a computational model that simulates the structure and functions of biological neural networks.[2] Figure 13-2 contains the basic categories, which we use to discuss the key differences of the human brain with an ANN.

| Categories of differences | Human brain | Artificial Neural Network (ANN) |
|---|---|---|
| Size | $10^{11}$ neurons <br> $10^{14}$–$10^{15}$ synapses (1–$10^5$ per neuron) | $16 \cdot 10^9$ neurons (size of a frog brain) <br> $10^9$–$10^{10}$ synapses |
| Topology | Complex, not sequential layers | Layer-based |
| Learning algorithm | We don't really know | Gradient descent with backpropagation |
| Power consumption and speed | Lower | Higher |
| Phases | Parallel | Training → prediction → evaluation |
| Execution | Asynchronously | Synchronously |
| Purpose and intent | Not static, can learn to learn | Use case / application specific |
| Memorization | Humans forget, less accurate over time | ANNs don't forget, always accurate |
| Cognitive capabilities | Innovation, creativity, desire, curiosity, understanding, inspiration, reasoning, character, intention, emotion | None to limited |
| Graphic | 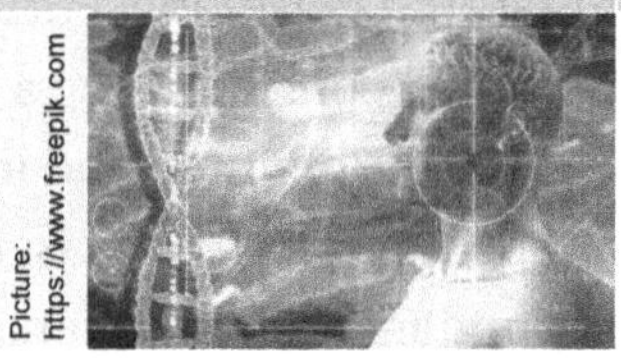 | 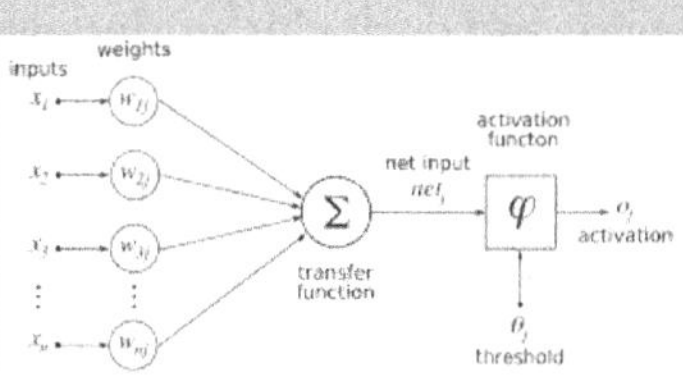 |

*Figure 13-2. The Human Brain and an Artificial Neural Network (ANN)*

[2]See [2] and [3] for more information on neural design, neural information theory, and the mathematical background on DL.

The human brain consists of roughly $10^{11}$ neurons and $10^{14}$ to $10^{15}$ synapses (with 1 to $10^5$ synapses per neuron), whereas the largest ANN has about $16 \bullet 10^6$ neurons and $10^9$ to $10^{10}$ synapses, which is approximately the size of a frog brain.[3] This comparison by itself illustrates the superiority of the human brain – at least as of today. In addition, the topology of the human brain is complex, *full of multidimensional geometrical structures operating in as many as 11 dimensions,*[4] not structured as sequential layers, like an ANN.

How our brain works is subject to research.[5] Much about this may be known; however, the way we learn is for all intents and purposes to a large degree still unknown to us. Whether it can be described as a set of algorithms, including the most sophisticated back-propagation algorithms of a DL model, is more than questionable. Learning of an ANN is primarily implemented through various back-propagation algorithms. The phases of learning as we know them from ML and DL models with training, validation, testing, prediction, retraining, and so on are performed differently by humans. An ANN has essentially a synchronous execution pattern, whereas the human brain works in a massively asynchronous, parallel fashion. For instance, visual recognition of the brain works asynchronously; the colors, form, and direction of movement of objects are processed and assembled asynchronously.

The objective or target of an ANN is specific to an application or scenario. For any ANN, there is a defined purpose. We don't have flexible multitask learning capabilities of ANNs (yet); no ANN can learn to learn for an arbitrary set of diverse circumstances or objectives. But the human brain can; it is adaptable. The human brain comes up with innovative ideas to even change its environment and the social and cultural codex in order to allow humans to survive. ANNs never forget anything;[6] they are always accurate in the context of their learned body of knowledge. Humans forget, and they are less accurate, which can be – under certain circumstances – an advantage.

Pervasive AI needs to include cognitive capabilities, such as innovation and creativity, desire and curiosity, understanding, and reasoning. Furthermore, it should include the ability to become inspired by something, or to develop a character (a unique, personalized behavior that is underpinned by a specific opinion and inhibition and rooted within a cultural context), to be driven by an intention, and – last but not least – it should develop and cope with emotions, not only understand sentiments from social media data. Situations where these cognitive capabilities are needed are definitely suggesting to us current *limitations of AI* or *how to use AI* in supporting human decisions.

---

[3]See [4] for more information on comparing biological with ANNs.

[4]See [5] for more information on the structure of the human brain.

[5]See [6] for more information on how our brain works.

[6]See the section on additional research topics on novel approaches for ANNs to forget.

# Current AI Limitations

AI is not comprised of limitless capabilities; there are noticeable limitations that have given rise to both skepticism and research as well. This section is devoted to the following AI limitations, which enables us to answer the question when not to use AI *yet*, or when AI is more an accompanying method:

- Labeling and annotation

- Autonomous ML and DL

- Multitask learning

- Explainability of decisions

Let us have a brief look at these four limitations of AI. We provide links for the interested reader to further study these areas in more depth.

# Labeling and Annotation

Labeling data is an essential ML and DL task. Unlabeled *raw* data that serves as input for training ML or DL models needs to be tagged with additional labels, which we could consider as metadata. For instance, adding classification information (e.g., fraud or non-fraud, churn or non-churn, accept or not accept of a marketing offer) transforms *raw* data to *labeled* data that serves as input for the learning (training) step of an ML classification model. Annotation is the process to label data, such as images, videos, audio, text, and others, to make the data usable for training primarily DL models.

The needs and challenges for labeling and annotation can differ by industry. For instance, the many different techniques and usage scenarios, such as bounding boxes around cars, motorbikes, traffic lights, and many more as part of video or image annotation for autonomous driving, or face detection and recognition of security surveillance cameras have one characteristic in common: they can require enormous human resources and very specialized tools and services and take up a lot of time.

Data labeling is indeed a key research area as well.[7] Issues are often addressed by applying other techniques, such as reinforcement learning and deep learning with limited or even zero input. These techniques, however, are essentially based on trial and error methods, which are impossible to be used for some scenarios, like autonomous driving or pattern recognition of complicated medical data, which may be used to recommend medication.

---

[7]See [7] and [8] for more information on data labeling for medical applications.

# Autonomous ML and DL

One of the biggest challenges for training ANNs is to obtain large enough data sets for training and to conduct the very time- and resource-consuming learning process that often involves human intervention. Autonomous ML and DL has the aspiration to simplify this process through continuous automation of learning, where self-governing processes with limited or even no human intervention enable the improvement and adaptation of an ANN. This is an emerging research area, which has the potential to develop ANN methods that can even cope with new circumstances and scenarios.

Today, there exist approaches to increase the flexibility of trained ANNs to adapt, for instance, the depth and structure of the ANN or to even develop an ANN *from scratch with the absence of an initial network structure via the self-constructing network structure.*[8] Similar objectives exist for reinforcement learning (RL), where an agent learns to improve its actions based on the rewards it receives. The challenge of autonomous RL includes to learn how to select relevant information out of data or image streams, where semantic, relevance, or meaning is not initially provided to the agent. The system may also be put *into an unrecoverable state from which no further learning is possible,*[9] which requires innovative approaches to reset systems for subsequent learning cycles. Autonomous DL includes learning with zero input. AlphaGo Zero[10] is a stunning success, where a combination of an advanced search tree and ANNs was developed, winning against world-class Go players.

Although these are promising results and research areas, the application to real-world scenarios, such as self-directing cars, industrial robotics, radiology or cancer diagnosis, surgical robotics, targeted treatment, and many others, is somewhat limited. In those scenarios, learning of ANNs or other models needs to be implemented prior to applying them in the field.

# Multitask Learning

In AI, learning is typically targeting a particular task or problem with a single KPI or measure. In real-world scenarios, however, learning should be multifold. This could, for instance, mean to not only recognize a face via a surveillance camera but to also interpret

---

[8]See [9] for more information on autonomous DL.

[9]See [10] for more information on autonomous RL.

[10]See [11] for more information on AlphaGo and AlphaGo Zero.

the facial expression regarding sentiments and emotions. Understanding speech should be complemented with recognizing the tonality, an accent, or the gender of the speaker. To train a DL model for related but different tasks, which is called multitask learning, is an emerging area of DL,[11] where the generalizability of learning should overcome the limitation to one use case or task only.

The human brain performs multitask learning in almost all situations, regardless whether we are looking at a picture in a museum, listening to another person during a presentation, or driving a car. As a car driver, for instance, when approaching an intersection, we are not only recognizing the traffic signs, a green light, or a car in front of us; we may look at and interpret the facial expression of another car driver approaching the intersection, or predict the likelihood of an approaching car still being able to properly stop at the intersection or not, or the likelihood of small children at the walk side crossing the street when they should rather not, or that the cyclist in front of us is an elderly person, and so forth. The majority of drivers have learned to factor in these circumstances, which may lead to a decision to reduce the speed further, be ready to brake, or to even give way despite having right of way.

These examples illustrate in a striking way that current *limitations of AI* requires us to determine *when* and *how* to use *which* AI capabilities and to what degree we can rely on AI supporting human decision making. To be cognizant about AI limitations determines the applicability of AI to certain problems and scenarios.

# Explainability of Decisions

The lack of explainability is not at all a new issue for AI systems. With the increasing degree of sophistication of ANNs or AI systems in general, explaining why and how an AI system, with its growing independent existence, induced a particular decision becomes increasingly important. Making AI-derived decisions comprehensible, interpretable, and trustable for a human being is of essence. We have touched on these topics already in Chapter 8, "*AI and Governance.*"

The rationale behind AI-based decisions needs to be made transparent to human beings. IBM Research, for instance, has developed a comprehensive strategy that "*addresses multiple dimensions of trust to enable AI solutions that inspire confidence,*"[12] including an AI Explainability 360 Open Source Toolkit for detecting, understanding, and mitigating unwanted algorithmic bias.

---

[11]See [12] for more information on multitask learning.

[12]See [13] for more information on IBM Research's strategy to enable AI solutions to inspire confidence.

# Insoluble Challenges

It must have become obvious to you that the human brain is and remains unsurpassed by AI. Of course, there are areas where AI is clearly superior; however, the human brain shines with its cognitive capabilities, its adaptability to new circumstances, and that it can cope with weird and odd situations. Despite stunning advances of AI and convincing research results, there are and remain – as we are convinced of – some insoluble AI challenges, which we elaborate on in this section.

## Cognitive Capabilities

As we have pointed out in the section *"AI and the Human Brain,"* the cognitive capabilities are certainly a key differentiator of the human brain and have always been a matter of fascination. In our discussion, we are focusing on those capabilities that may never be attained by AI without leaving noticeable and major gaps.

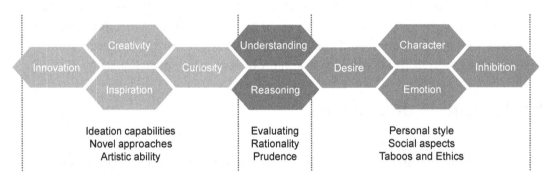

*Figure 13-3.* *Cognitive Capabilities*

As you can see in Figure 13-3, we have categorized the cognitive capabilities of the human being into three groups. Human beings – at least most of them – are driven by curiosity and inspiration. We do not only realize and implement instructions given to us by others; we develop our own ideas and are looking for novel approaches. Innovation and creativity are synonyms for being human. Could the most powerful ANN even *ask the right questions* that led to the development of Einstein's theory of general or special relativity? Could the concept of democracy be invented by AI, and could an ANN invent a completely new style of music, not copying what others have done or copying a composer's style, but creating something completely new? This is where we as human beings scintillate – and always will; AI can be a useful companion, not more.

Decisions need to be evaluated and implemented by exercising prudence. The rationality of the decision-making process is often as important as the decisions themselves. Understanding and reasoning are aspects that should complement the plain decisions. Scrutinizing the meaningfulness and reasonability of decisions within a given context and circumstances is something AI systems are struggling with. Again, AI can support you in numerous scenarios and tasks, but understanding and reasoning are eventually left to you.

Every person is unique; we develop a particular character and personal style in addressing problems. We have desires and develop emotions; our decisions, activities, and business practices are well rooted within a social and cultural codex, where taboos and ethical aspects suggest boundaries and often even an inhibition to do certain things.

These sets of cognitive capabilities make our interconnected way of working together and dealing with each other in colorful and interesting ways. Tolerance, consideration for others with other opinions and attitudes, emotions and humor, and the ability to restrain oneself and compromise were needed; these are cognitive behavioral skills that are needed in many situations, where AI is certainly not in the driver seat – and may never be.

Not learning from data and the environment, but learning *how to change* the environment seems to be a level of human intelligence, which is unsurpassed by AI.

## Weird Situations

Neither can all real-life situations be predicted with sufficient accuracy, nor is it possible to determine adequate, human-equivalent reactions to those situations. We are illustrating this using autonomous driving as an example.

Think about situations – as seldom as the may occur – that a self-driving car might have to cope with: a truck or a bicycle coming along the wrong way, an ambulance vehicle making an illegal turn, children crossing the street while cars have green light, or a bus driver giving a hand signal to warn approaching cars of a hazardous situation in front of them.

Even if autonomous driving systems will be able to detect these situations, they may not be able to adequately react as a human driver would have learned to.[13] Autonomous driving systems with their radar equipment, cameras, and sensors may detect other cars and bicycles, traffic signs, and objects on the street; however, predicting behavior of other road users, pedestrians, or even animals and dealing with the unexpected may never be satisfactorily possible for any AI system.

---

[13]See [14] for more challenges regarding self driving cars in weird situations.

# Generalization of Learning

One particular aspect that may be holding back your AI endeavors is the missing adaptation of AI and ANNs to completely new circumstances. The generalization of learning, which allows adaptation and learning of an arbitrary set of different disciplines and scenarios, is simply impossible today.

This aspect reaches even far beyond multitask or autonomous learning concepts, which we have discussed earlier in this chapter. In general, ML and DL models remain applicable for a given set of initial scenarios or problems. Once trained, those models achieve fascinating performance; however, the underlying model – even an ANN – will struggle to be applicable for another scenario or domain with a different set of untrained data.

AI research has made great progress in regard to generalizing learning[14]. It may sound strange, but an important aspect of generalizing learning for ANNs is to be able to forget past experiences, for instance, deadlock situations or catastrophic events.

Despite promising results, the maturity of these approaches and applicability for real-life business problems is still in the distant future.

# Additional Research Topics

As you can imagine, there are quite a number of additional AI research topics, which could easily fill a book by itself. The following Table 13-1 contains a very short description of some research areas, which further illustrate the limitation of today's AI capabilities. Although it may not be required for you to dive into the details, but our readers should be aware of those areas in regard to understanding the *limitations of AI* and *how to use AI* properly.

---

[14]See [15] for more information on human-like generalization of learning.

***Table 13-1.*** *AI Research Areas*

| # | Research area | Description |
|---|---|---|
| 1 | Hyperparameter optimization (HPO) | Learning is done via algorithms. Optimizing and tuning the learning algorithm is done via HPO, for example, the number of ANN layers, number of decision trees and their depth, the learning rate, and so on. AI research is concerned about automating and accelerating HPO approaches, to eliminate or at least assist manual tuning |
| 2 | Back-propagation algorithms | In essence, back-propagation algorithms are used to learn and improve an ANN by adjusting weights. There are research efforts to accelerate those algorithms (e.g., via on-chip acceleration). In addition, research is focusing on alternative approaches, moving beyond supervised DL with back-propagation (e.g., autoencoders, adversarial networks, neuroevolution that uses evolutionary algorithms, etc.) |
| 3 | Learning to learn | Humans can learn how to learn in order to approach and eventually solve future unknown problems. In AI, this is addressed by an attempt to develop meta-learning approaches and to automate the design of ML and DL models. This goes beyond the generalization of learning, requiring flexibility, common sense, human learning approaches, and even human instincts, which humans have developed over millions of years |

There are additional AI research areas,[15] not listed in the table, related to adversarial networks,[16] RL, conversational systems, specialized hardware for AI, and speech, just to name a few.

# Key Takeaways

We conclude this chapter with a few key takeaways, summarized in Table 13-2.

---

[15]See [16] for more AI research areas.

[16]Adversarial networks are ANN network architectures, where two ANNs compete and work with each other to improve the overall accuracy of the resulting ANN.

*Table 13-2.*  *Key Takeaways*

| # | Key Takeaway | High-Level Description |
|---|---|---|
| 1 | Human brain compared to ANN | There are key differences between the human brain and ANNs, in terms of size, topology, capabilities, learning approaches, and so on |
| 2 | There are currently AI limitations | There are a number of AI limitations, such as labeling and annotation, autonomous AI, multitask learning, and explainability of decisions |
| 3 | ANNs don't have key cognitive capabilities | Cognitive capabilities (creativity, inspiration, curiosity, emotions, understanding, reasoning, desire, etc.) of the human brain are a key differentiator to ANNs |
| 4 | There are insoluble AI challenges | There are a number of insoluble challenges for AI, such as cognitive capabilities, coping with weird situations, and generalizing learning |
| 5 | AI research topics | There are quite a number of AI research topics that are geared toward automating and accelerating learning and development of ML and DL models |
| 6 | Learning to learn | Humans can learn how to learn to address and solve future unknown problems and challenges; AI is still struggling with this |

# References

[1]  Fillard, J.-P. *Brain Vs Computer: The Challenge Of The Century.* ISBN-13: 978-9813145542, World Scientific, 2016.

[2]  Stone, J.V. *Principles of Neural Information Theory: Computational Neuroscience and Metabolic Efficiency (Tutorial Introductions).* ISBN-13: 978-0993367922, Sebtel Press, 2018.

[3]  Stone, J.V. *Artificial Intelligence Engines: A Tutorial Introduction to the Mathematics of Deep Learning.* ISBN-13: 978-0956372819, Sebtel Press, 2019.

[4]   PHYS.ORG. Eindhoven University of Technology. *New AI method
      increases the power of artificial neural networks*, 2018, `https://
      phys.org/news/2018-06-ai-method-power-artificial-neural.
      html` (accessed October 18, 2019).

[5]   Dean, S. Science Alert. *The Human Brain Can Create Structures
      in Up to 11 Dimensions*, 2018, `www.sciencealert.com/science-
      discovers-human-brain-works-up-to-11-dimensions` (accessed
      October 18, 2019).

[6]   Sheehan, T.D. *The Oscillating Brain: How Our Brain Works*. ISBN-
      13: 978-1489705815, LifeRich Publishing, 2016.

[7]   Carneiro, G. et.al. *Deep Learning and Data Labeling for Medical
      Applications* (Lecture Notes in Computer Science, Band 10008).
      ISBN-13: 978-3319469751, Springer, 2016.

[8]   Landgraf, M. Karlsruhe Institut of Technology. *Training Data for
      Autonomous Driving*, `www.kit.edu/downloads/pi/PI_2019_048_
      Training%20Data%20for%20Autonomous%20Driving.pdf`
      (accessed October 20, 2019).

[9]   Ashfahani, A., Pratama, M. *Autonomous Deep Learning: Continual
      Learning Approach for Dynamic Environments*, `https://arxiv.
      org/pdf/1810.07348.pdf` (accessed October 21, 2019).

[10]  Eysenbach, B., Gu, S., Ibarz, J., Levin, S. *Leave no Trace: Learning
      to reset for safe and Autonomous Reinforcement Learning*,
      `https://openreview.net/pdf?id=S1vuO-bCW` (accessed October
      22, 2019).

[11]  Silver, D., Hassabis, D. *DeepMind. Research Blog Post. AlphaGo
      Zero: Starting from scratch*, `https://deepmind.com/blog/article/
      alphago-zero-starting-scratch` (accessed October 22, 2019).

[12]  Ruder, S. *An Overview of Multi-Task Learning in Deep Neural
      Networks*, `http://ruder.io/multi-task/` (accessed October 22,
      2019).

[13]   IBM. *Trusting AI– IBM Research is building and enabling AI solutions people can trust,* `www.research.ibm.com/artificial-intelligence/trusted-ai/` (accessed October 23, 2019).

[14]   The New York Times. *Despite High Hopes, Self-Driving Cars Are 'Way in the Future',* `www.nytimes.com/2019/07/17/business/self-driving-autonomous-cars.html` (accessed October 25, 2019).

[15]   Doumas, L.A.A., Puebla, G., Martin, A.E. Human-like generalization in a machine through predicate learning, `https://arxiv.org/ftp/arxiv/papers/1806/1806.01709.pdf` (accessed October 26, 2019).

[16]   IBM. AI Research, `www.research.ibm.com/artificial-intelligence/` (accessed October 27, 2019).

# CHAPTER 14

# In Summary and Onward

In this book, we explained how AI can today be used in enterprises. We dove into key aspects like an *AI information architecture (IA)* to lay a foundation of data in order to support AI, the *AI life cycle* to get from data to predictions to optimal decisions and actions, and important *AI operations (AIOps)* and *AI DevOps* aspects. We have furthermore elaborated on additional enterprise aspects, such as *AI deployment and operationalization* challenges, AI in the context of *governance*, *change management*, *design thinking*, and *MDM*. We have also exposed you to some limitations of AI – including limitations that may persist for the foreseeable future – and some exciting and emerging topics, such as AI in the context of *blockchain* and *quantum computing*.

In the final chapter of this book, we want to look ahead and share some thoughts on how AI, the application of AI in enterprises, and ultimately new enterprises and industries built on AI will drive profound changes in the future.

## AI for the Enterprise – Low Hanging Fruit

As discussed through this book, already today AI brings a lot of value for enterprises and organizations large and small who are ready and able to adopt it. There is plentiful opportunity to go further in making enterprises, organizations, and even small companies faster and more efficient, by augmenting or automating an increasing number of everyday decisions using AI, in order to improve customer satisfaction, reduce cost, and optimize staffing.

AI skills are becoming more prevalent, while at the same time developing AI solutions becomes quicker and easier enabled by better tools and readily available compute capacity to power AI on public and private clouds. Using state-of-the-art tools and popular cloud platforms, even small companies and individual lines of business can benefit from AI, without owning a single server and without employing a single IT administrator. Skills, tools, and pervasively available cloud compute capacity without the need for own IT will lead to further acceleration in industrialization and broad-scale adoption of AI.

© Eberhard Hechler, Martin Oberhofer, Thomas Schaeck 2020
E. Hechler et al., *Deploying AI in the Enterprise*, https://doi.org/10.1007/978-1-4842-6206-1_14

It is hard to imagine how much today's enterprises can be streamlined, optimized, and accelerated once they leverage all available aspects of AI end to end across their entire value chain and in all interactions with users, customers, partners, and suppliers. It is well possible that some enterprises that really invest in and focus on AI over the next few years can achieve KPIs such as two times faster and better service for their customers with 25% less staff and 50% less expenses.

# The AI Enterprise – Whitespace

The potential of AI for the enterprise is enormous. However, constraining our forward thinking exclusively to using AI for the enterprise would be limiting. Applying AI in more and more businesses and processes of existing enterprises will improve efficiency and speed, but still will merely result in a better version of the same enterprise.

We are convinced that *the full potential of AI will only be achieved by inventing new **AI enterprises***. What this means in practice will be very different in each individual case. Founders of new enterprises and leaders of existing ones need to think fresh:

- Considering what AI can do – not only today but over the lifetime of the enterprise – what is the enterprise's fundamental purpose and value?

- How does the enterprise have to be structured to achieve that purpose and provide sustained value to customers and users?

- How does it need to interface with users, customers, suppliers, and partners in order to execute and operate best in the context of that structure?

These are just three of many questions for which new AI enterprises should have novel answers. Inventing an AI enterprise is not mainly a technical challenge; it requires to see entirely new opportunities to provide customers with something that they do not yet know they want or need, but becomes essential to them once they have it. Let's illustrate this via an example.

# An Example

The purpose of a new AI enterprise can be radically different from the purpose of a classic enterprise – it does not have to be constrained by existing concepts, business models, ecosystems, markets, customers, or culture.

A newspaper – even in the age of the Internet – might have defined its purpose as providing its readers with important information and facts through articles. To achieve that purpose, they may have chosen a structure of employing reporters and editors to write and curate articles, keeping and acquiring readers and advertisers who pay for ads. They may own buildings, offices, printing machines, and so on. The interface to readers may be a website and mobile application showing articles to readers and allowing readers to like articles and comment. There may still also be a printed version for readers who still want to read the news on paper.

In contrast, a new AI enterprise may invent its purpose fundamentally differently, as engaging its users in a continuous multidirectional exchange of information, showing them information they are interested in in real time, as well as allowing them to contribute. To achieve that different purpose, the AI enterprise may invent its structure to be purely web based, with an app and website fronting an AI-powered information exchange platform on a cloud for billions of users. The interface to users may be multimodal, showing, reading, or displaying to each individual user what the AI-powered platform learned that particular user is interested in and allowing each user to record or write and submit their own contributions if they so choose, which then in turn will be targeted to those other users interested in that information. We all know enterprises in this space that have grown from startups to large enterprises with billions of users in a period of a few years.

As in the preceding example, an AI enterprise does not need to adopt the same purpose, structure, and interfaces for customers, users, partners, and suppliers as an existing enterprise in a space or domain would. It does not have to constrain itself to merely apply, for instance, ML and DO to optimize its processes, operations, cost, and staffing. Instead, an AI enterprise can choose a new and different purpose that would never be possible to achieve without AI, a completely different structure and business model to achieve that purpose, and different interfaces to engage users and customers.

# Future of AI

Thinking and inventing new big ideas outside known paradigms is nicely expressed by Henry Ford's statement *"If I had asked people what they wanted, they would have said faster horses...."* There is a lot of truth and wisdom in this statement.

AI brings about a fundamental shift in the art of the possible. In order to unleash its full potential, it will be critical to not merely use AI "to make a faster horse" and streamline and accelerate existing enterprises, but to invent new AI enterprises and to reinvent existing enterprises with new purpose that would be impossible without AI.

This journey starts by applying AI to your enterprise with its current business scope and models. This is what we have tried to focus on in this book. There is virtually no enterprise that would not be able to benefit from utilizing AI.

However, this is just the beginning: AI offers you its full breadths of potential once you are realizing a paradigm shift, thinking about AI in an unbiased and entrepreneurial way.

*Applying AI to make existing enterprises better is inevitable. But its true potential is in enabling creation of entirely new kinds of enterprises that could never exist without AI.*

# CHAPTER 15

# Abbreviations

| | |
|---|---|
| ABB | Architecture building blocks |
| AGI | Artificial general intelligence |
| AI | Artificial intelligence |
| AIaaS | AI as a service |
| AIIA | AI information architecture |
| AIIRA | AI information reference architecture |
| AIOps | AI for IT operations |
| AIRA | AI reference architecture |
| ANN | Artificial neural network |
| AOD | Architecture overview diagram |
| API | Application programming interface |
| AQL | Annotation Query Language |
| ARM | Advanced RISC machines |
| ASIC | Application-specific integrated circuit |
| ATM | Automated teller machine |
| AWS | Amazon Web Services |
| BCBS 239 | Basel Committee on Banking Supervision Standard 239 |
| BCO | Beneficial cargo owners |

(*continued*)

© Eberhard Hechler, Martin Oberhofer, Thomas Schaeck 2020
E. Hechler et al., *Deploying AI in the Enterprise*, https://doi.org/10.1007/978-1-4842-6206-1_15

| | |
|---|---|
| BNN | Bayesian belief network |
| CADS | Cognitive assistant for data scientists |
| CC | Cognitive computing |
| CCO | Chief compliance officer |
| CCPA | California Consumer Privacy Act |
| CDA | Chief data officer |
| CDI | Customer data integration |
| CDP | Customer data platform |
| CICS | Customer information control system |
| CISO | Chief information security officer |
| CI/CD | Continuous integration/continuous delivery |
| CNN | Convolution neural networks |
| COBIT | Control Objectives for Information and Related Technologies |
| COV | Covariance |
| CR | Change request |
| CRUD | Create, read, update, and delete |
| CT Scan | Computed tomography scan |
| DBM | Deep Boltzmann machines |
| DBMS | Database management system |
| DBN | Deep belief networks |
| DL | Deep learning |
| DLN | Deep learning networks |
| DNA | Deoxyribonucleic acid |
| DO | Decision optimization |
| DOB | Date of birth |

*(continued)*

| | |
|---|---|
| DPO | Data protection officer |
| DR | Disaster recovery |
| DRL | Deep reinforcement learning |
| DWH | Data warehouse |
| EDI | Electronic data interchange |
| EIA | Enterprise information architecture |
| EIARA | Enterprise information architecture reference architecture |
| EMR | Elastic MapReduce |
| ETL | Extract, transform, load |
| FDA | Flexible discriminant analysis |
| FPGA | Field programmable gate array |
| GCP | Google Cloud Platform |
| GDPR | General Data Protection Regulation |
| GNN | Graph neural network |
| GPU | Graphics processing unit |
| GRC | Governance, risk, and compliance |
| GUI | Graphical user interface |
| HA | High availability |
| HIPAA | Health Insurance Portability and Accountability Act |
| HPC | High-performance computing |
| HPO | Hyperparameter optimization |
| IA | Information architecture |
| IaaS | Infrastructure as a service |
| IoT | Internet of Things |
| ISACA | Information Systems Audit and Control Association |
| ISV | Independent software vendor |

(*continued*)

| | |
|---|---|
| ITIL | Information Technology Infrastructure Library |
| ITSM | IT service management |
| JDBC | Java Database Connectivity |
| JSON | JavaScript Object Notation |
| KPI | Key performance indicator |
| LDA | Linear discriminant analysis |
| MDA | Mixture discriminant analysis |
| MDM | Master data management |
| MDP | Markov decision process |
| ML | Machine learning |
| MLP | Multilayer perceptron |
| MPP | Massively parallel processing |
| MRI | Magnetic resonance imaging |
| MRO | Maintenance, repair, and operating |
| NLP | Natural language processing |
| ONNX | Open Neural Network Exchange |
| PaaS | Platform as a service |
| PBFT | Practical Byzantine Fault Tolerance |
| PC | Principal component |
| PCA | Principal component analysis |
| PFA | Portable format for analytics |
| PII | Personal identifiable information |
| PIM | Product information management |
| PLSR | Partial least squares regression |
| PMI | Project Management Institute |
| PMML | Predictive Model Markup Language |
| POPI | Protection of Personal Information Act |

(continued)

| | |
|---|---|
| POS | Point of sale |
| PR | Precision-recall |
| Qubit | Quantum bit |
| QLDB | Quantum Ledger Database |
| QRAM | Quantum random access memory |
| RBFN | Radial basis function network |
| RDM | Reference data management |
| RL | Reinforcement learning |
| RNN | Recurrent neural networks |
| ROC | Receiver operating characteristic |
| SaaS | Software as a service |
| SEC | Securities and Exchange Commission |
| SHRM | Society for Human Resource Management |
| SMP | Symmetric multiprocessing |
| SOC | Security operations center |
| SOX | Sarbanes-Oxley Act |
| SSL | Secure Socket Layer |
| SSN | Social security number |
| SSO | Single sign-on |
| SVM | Support vector machines |
| TOGAF | The Open Group Architecture Framework |
| TPU | Tensor processing unit |
| UX | User experience |
| VAR | Variance |
| VM | Virtual machine |

# Index

## A

Adapting AI models, 126, 127
Aerospace industry, 29
AI as a service (AIaaS), 72
AI guarding agent, 197
AI information architecture (AIIA), 59
AI information reference architecture
(AIIRA), 59
AI model monitoring
bias and fairness, 134
outcome, 134
performance, 133, 134
AI operations (AIOps), 313
AI reference architecture (AIRA), 58
Amazon Kinesis Data Firehose, 83
Amazon Kinesis Video Streams, 83
Amazon Web Services (AWS)
AI information architecture overview
diagram, 83
analytics services, 83
governance, 201
ML services and tools, 84
Apache Spark models, 129
Application integration, 131
Architecture building blocks (ABBs), 66
information governance and
information catalog layer, 195
Architecture Development Method
(ADM), 59
Architecture overview diagram (AOD), 67

AI information architecture, 68
building blocks, 65, 67
Artificial general intelligence (AGI), 3
Artificial intelligence (AI)
automated decisions (*see* Automated
decisions)
enterprise, 4, 5
IA (*see* Information architecture (IA))
Artificial neural networks (ANNs), 9, 24,
45–47, 301
Assets, 253
Auto AI
feature importance analysis, 44
model training pipelines, 44
predictive ML models, train, 101–103
set of models, 43
Automated decisions
actions require decisions, 5, 6
decisions require predictions, 6, 7
prediction and optimization, 8
Automotive industry, 29
Autonomous learning, 299
Autonomous ML and DL, 304

## B

Back-propagation, 47
Batch scoring, 130
Bayesian belief network (BNN), 40
Bayesian network, 40
Biased data, 10